Brian Stefaniak

Cosmic Pinball

The Science of Comets, Meteors, and Asteroids

Cosmic Pinball

The Science of Comets, Meteors, and Asteroids

BY CAROLYN SUMNERS
AND CARLTON ALLEN

McGraw-Hill

New York San Francisco Washington, D.C. Auckland Bogotá
Caracas Lisbon London Madrid Mexico City Milan
Montreal New Delhi San Juan Singapore
Sydney Tokyo Toronto

McGraw-Hill

A Division of The ***McGraw-Hill*** *Companies*

pbk 1 2 3 4 5 6 7 8 9 0 AGM/AGM 9 0 9 8 7 6 5 4 3 2 1 0 9

ISBN 0-07-135460-3

Printed and bound by Quebecor/Martinsburg.

Project supervision: North Market Street Graphics
Title page illustration courtesy of Pat Rawlings (copyright © 1999 Pat Rawlings); Part 1 photo courtesy of the Association of Universities for Research in Astronomy, (AURA); Part 2 photo courtesy of B. Wilson; Part 3 photo courtesy of Carl Allen; Part 4 illustration courtesy of Dan Durda; Part 5 photo courtesy of Gary Young and Carolyn Sumners.

For Gene Shoemaker, who saw so clearly.

CONTENTS

Part One: The Cosmic Pinball Risk
Carolyn Sumners and Carlton Allen

INTRODUCTION **The Cosmic Pinball Risk** 3

Part Two: Comets
Carolyn Sumners

CHAPTER 1 **The Comet Connection** 11

CHAPTER 2 **The Cast of Comet Characters** 29

CHAPTER 3 **Close Comet Encounters** 47

Part Three: Meteors and Meteorites
Guy Consolmagno and David Kring

CHAPTER 4 **A Rain of Comet Dust** 65

CHAPTER 5 **Meteorite Hunting** 81

CHAPTER 6 **Rocks from the Moon and Mars** 99

Part Four: Asteroids
Dan Durda and David Kring

CHAPTER 7 **Asteroids in Orbit** 119

CHAPTER 8 **Asteroid Spacecraft Encounters** 133

CHAPTER 9 **Doomsday Asteroids** 149

Part Five: Assessing the Risk
David Kring

EPILOGUE **The Threat of Future Impacts on Earth** 169

INDEX 179

Thanks first and foremost to our authors, leaders in the new field of planetary science. Dr. Guy Consolmagno is a Jesuit brother on the staff of the Vatican Observatory. Guy specializes in the complex relationships between meteorites and asteroids, and between faith and science. Dr. Dan Durda is a postdoctoral researcher at the Southwest Research Institute in Boulder, Colorado. Dan is a talented space artist whose research concentrates on the collisional histories of asteroids. Dr. David Kring is a member of the research staff of the Lunar and Planetary Laboratory at the University of Arizona. David studies the effects of asteroid impacts and was one of the discoverers of the Chicxulub crater, probable ground zero for the extinction of the dinosaurs.

Special thanks also to our team of editors. Jaclyn Allen, geologist and science educator from Lockheed Martin at the Johnson Space Center, has worked many long hours to ensure that the text is both readable and accurate. M. R. Carey of North Market Street Graphics has made sure that the English is proper and the details are right. Our McGraw-Hill Editor, Griffin Hansbury, recognized the need for this book. His patience and support have been vital to the successful completion of *Cosmic Pinball.*

Carolyn Sumners
Director of Astronomy
Houston Museum of Natural Science

Carlton Allen
Planetary Scientist
Lockheed Martin/Johnson Space Center

PART ONE

The Cosmic Pinball Risk

The Cosmic Pinball Risk

***COSMIC PINBALL* IS A METAPHOR FOR THE ACTIVITY IN EARTH'S NEIGH**borhood viewed on a cosmic time scale. The projectiles—comets, meteorites, and asteroids—have already been launched. Jupiter's gravity, like giant pinball flippers, continues to bounce these objects back into the inner solar system where Earth waits. A cosmic pinball game requires thousands or even millions of years to play, but eventually the planet posts are hit. The pinball analogy breaks down when an object actually collides with a planet: There is no elastic bounce. The impacting object is usually destroyed, and the energy released at impact can change the planet forever.

This book contains the latest research and summarizes the risks associated with living in our dangerous cosmic neighborhood.

Wake-Up Calls

Racing through space at 74,000 kph, an asteroid crossed Earth's orbit on March 23, 1989—at a point just 650,000 km in front of the planet. This flying mountain, 400 m wide, passed only 170,000 km beyond the Moon.

A collision with Earth would have been catastrophic. Tidal waves 100 m high could have wiped out coastal areas. A direct hit would have obliterated a city the size of New York. Yet the asteroid passed unobserved. Astronomers discovered it as it rushed away from Earth.

Many times in the past, Earth has not been so lucky. Although no one chronicled the Chicxulub impact 65 million years ago, we can describe a likely scenario. The planet's atmosphere did little to slow down the 10-km-wide asteroid. Diving toward the Gulf of Mexico at over 80,000 kph, the trillion-ton intruder heated the atmosphere thousands of degrees. As the huge rock struck the ocean, the impact shock boiled trillions of tons of seawater and sent jets of vapor rocketing skyward. When the asteroid crashed into the seafloor, its impact released 100 million megatons of energy and shook the entire planet.

With a thunderous roar and blinding flash, over 1000 trillion tons of Earth's bedrock and seawater rushed from the impact site—leveling the landscape in a 1500-km radius. Shock waves from the explosion generated searing winds that scorched the planet.

Meanwhile, the crust rebounded with earthquakes, and a colossal wall of water rushed outward at 700 kph. Waves rolled inland, devastating much of what is now Mexico, Florida, and the U.S. Gulf Coast.

Dust ejected into space fell back over the planet—heating the air and igniting the land below. It is thought that as much as 90 percent of Earth's forests and grasslands may have burned.

Soot from fires combined with spreading dust formed a shroud greater than 20 km thick that kept out sunlight for as long as 6 months. Photosynthesis stopped and food-producing plants on land and in the oceans died. The food chain collapsed.

A deadly acid rain began to fall and the surface temperature plunged to below freezing over the darkened land. Many species (including dinosaurs), accustomed to warmer temperatures and abundant food, began to die. When the dust clouds finally lifted, a very different population of plants and animals had survived to reclaim the planet. Extinction of the great reptiles cleared the way for the rise of mammals on Earth.

Identifying the Risk

These stories and the associated risks are real. Through television docudramas, graphic magazine articles, and blockbuster movies, collision scenarios have been graphically illustrated. But often the disaster is presented without an assessment of how likely it is to occur. How often has Earth been hit, how many projectiles are in our neighborhood now, and when can we expect another disaster? Lacking any quantification of the risk, public reactions range from a little panic to a lot of indifference.

Yet proof of celestial target practice is everywhere—from moon craters to circular scars still visible on Earth. The threats are still out there somewhere, but where are they, how many are there, what are their paths, and how can we protect ourselves? These questions form the focus of this book.

Impact hazards come in three classes—comets, meteorites, and asteroids—each with a unique appearance, history, and associated risk. Because of their ghostly appearance, unusual paths, and unpredictability, icy comets have the worst reputation of our wandering neighbors. Comet phobia has been with us for thousands of years, yet no human death has ever been directly linked to a comet impact. Comets, however, may indeed pose the greatest long-term threat since they can appear from the outer reaches of the solar system without warning or sufficient time for preparation. Chapters 1, 2, and 3 deal with these notorious visitors and their past, present, and future roles in doomsday fact and fiction.

Meteorites are the smallest and most frequent visitors. They're often seen burning up in Earth's atmosphere as meteors or "shooting stars" and pose little threat to humans on the surface. Yet a collision with a spacecraft could be deadly—a very real threat for orbiting astronauts and bases on the airless moon. Larger objects which reach Earth's surface provide our only physical laboratory samples of the flying projectiles in our neighborhood. Chapters 4, 5, and 6 focus on meteor showers and meteorites.

Rocky asteroids pose the most immediate and significant risk for the earth's surface. The first asteroid was discovered in 1801, long after the first comets and meteor showers had brought fear to wary humans. Asteroids, even when large enough to do significant damage, are still very hard to detect and track. There are too many of them and too much sky to cover. Chapters 7, 8, 9, and the epilogue address these nearby neighbors and their ongoing threat.

Describing the Threat

Humans have developed a lingo for describing natural disasters. Earthquakes have the Richter scale; hurricanes and tornadoes are described by class. These numbers measure the energy stored in a given phenomenon and indicate its power to cause catastrophic damage.

A similar scale can be developed for comets, meteorites, and asteroids based on the anticipated effects of their impacts with Earth. The amount of potential destruction depends on the object's energy at impact. This *kinetic energy* is a function of the object's mass and the square of its velocity. An object's mass depends on the cube of its average diameter and its density. Both velocity and diameter can be measured and either could be used for the scale. An object's diameter is the observable characteristic that has the greatest effect on the energy of an impact and the greatest range of values.

Table I.1 Impact Effects

SIZE	IMPACT FREQUENCY	EFFECT AND REFERENCES
1 mm–1 cm (sand grain/pebble)	1 per second (thousands per day)	Bright "shooting star," destroyed in the atmosphere. Chapters 1, 3, and 4.
1 cm–0.5 m (rock)	1 per hour (more than 10 per day)	Fireball, most destroyed in the atmosphere. Chapters 4 and 5.
0.5–1 m (microwave oven)	1 per day	Bolide (brilliant fireball), most destroyed in the atmosphere. Chapter 5.
1–10 m (automobile)	1 per 10 years	Stony or icy boulders can be destroyed in the atmosphere; iron boulders and some others reach the surface and can crash through a roof or damage a car. Chapters 5 and 7.
10–50 m (house)	1 per 100 years	Local disaster, equivalent to several Hiroshima-sized bombs. Chapter 5.
50–100 m (football field)	1–2 per 1000 years	Regional disaster, equivalent to the Meteor Crater or Tunguska event (about 15 megatons of TNT). Chapters 2, 3, and 9.
100 m–1 km (small village)	1 per 50,000–500,000 years	Continent-size disaster, equivalent to thousands of megatons of TNT. Chapters 2, 3, and 5.
1–10 km (small city)	1 per 10–100 million years	Mass extinction, threat to all life (millions of megatons of TNT). Chapter 9.
>10 km	<1 per billion years	Threat to the continued habitability of the planet. Chapter 9.

A classification system based on size can be used for comets, meteorites, and asteroids. Comets are made of lighter materials than meteorites and asteroids and pack less punch for their size. But comets travel more than twice as fast as asteroids because of their elongated orbits. A lighter object traveling faster can be as deadly as a heavier, but slower moving, projectile. Therefore comets, meteorites, and asteroids of the same general size deliver approximately the same kinetic energy. The object's size, more than any other factor, determines the risk associated with an impact (Table I.1).

The chapters which follow address Earth's real and future danger from each kind of cosmic intruder: the comets, the meteorites, and the asteroids.

PART TWO

Comets

The Comet Connection

In the Beginning

Comets are ancient material left over from a time 4.6 billion years ago when the Sun, planets, and moons formed from a cloud of dust and gas. This cloud contained hydrogen, helium, and other gases, along with a rich mixture of ice and dust grains that furnished all the remaining elements now found in the solar system.

The universe is far older than our Sun. Generations of stars had lived and died before the Sun was born. Pressure waves from the explosive death of an ancient star might have triggered the collapse of the dust cloud and led to the formation of the Sun and its solar system. The abundance of different elements found in meteorites suggests that the explosions of ancient stars contributed material to the primordial solar cloud.

This cloud collapsed under gravity's pull to become a swirling disk of gas, ice, and dust. The central portion of the solar system birth cloud condensed into a new star, which began to fuse hydrogen into helium in its core. Radiation produced by this young star changed the rest of the cloud forever.

Within the cloud, swarms of stony and icy clumps grew, collided, broke up, and gradually collected together. Rocky lumps resembling stony asteroids and icy clumps resembling comets were the building blocks of the planets. Objects forming in the inner solar system near the hot new Sun became rocky as heat from the newborn star vaporized any icy component. Farther from the Sun, growing objects incorporated increasing amounts of ice.

As each planet grew larger, it became a bigger target with a stronger gravitational pull in this early game of cosmic pinball. Young planets attracted new materials, which would either stick or slingshot around the planet in a new trajectory. Bodies in the cooler outer solar system grew very large. Jupiter and Saturn, largest of the newborn planets, collected most of the gaseous and icy bodies, leaving some residue for the outer gas giants, Uranus and Neptune (Fig. 1.1).

Comets probably formed in the region where Uranus and Neptune were born. Here the physical conditions were right for icy bodies to

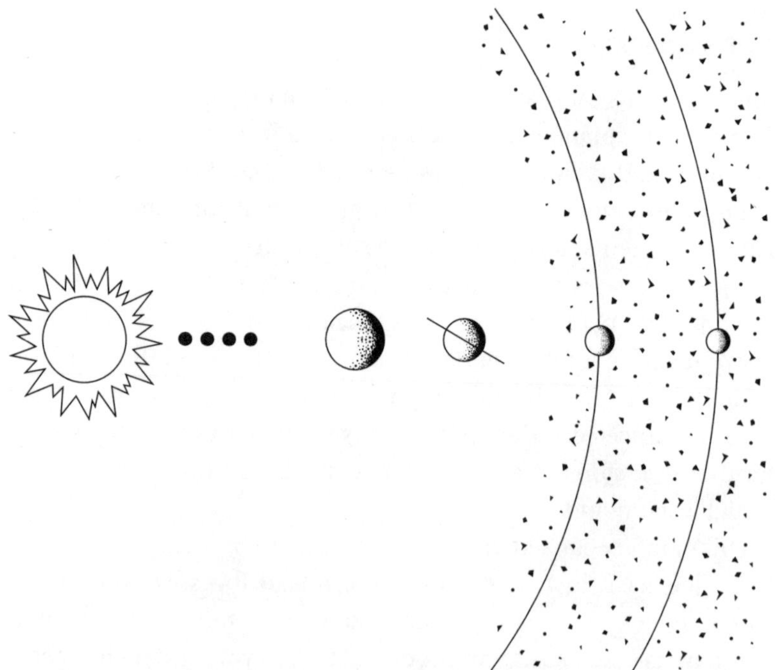

Figure 1.1 Orbits of planets. This scale drawing shows the large relative distances to the giant planets and the region of comet formation.

coalesce. A comet's nucleus is mostly water, which can turn into icy rock in cold areas far from the Sun. Kilometer-sized snowballs filled with ice and dust probably grew up near the massive giant planets. Over time, the planets disturbed the orbits of these young comets—sending some sunward and flinging many more into distant orbits at the edge of the solar system in a region called the *Oort cloud*, named in honor of Ian Oort, the Dutch astronomer who first proposed its existence. Eventually, many of the comets crossing into the inner solar system collided with the newly formed planets.

After planet formation had ended, mopping-up operations followed for millions of years. Planet surfaces collected chunks of ice and rock in violent cratering impacts. On Earth, the evidence is hidden; but other bodies, with much more ancient surfaces, display their battle scars prominently—just look at the Moon and Mercury. Some of the projectiles were comets, bringing ices to these rocky worlds. Comet ices trapped long ago in deep sunless craters at the Moon's poles may still be there (Fig. 1.2). This valuable resource could someday provide the water needed for a lunar colony.

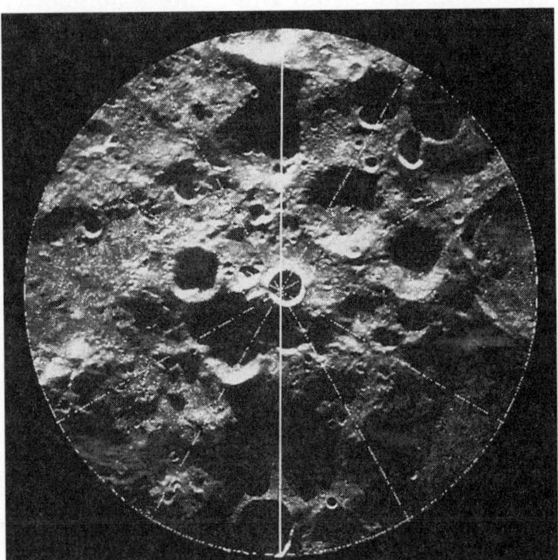

Figure 1.2 South pole of the Moon. Convergence of longitude lines shows the location of the Moon's south pole. Large parts of this area are permanently shadowed and could contain deposits of water ice. (Image courtesy of Lunar and Planetary Institute)

Comets are relics from the time of the formation of the solar system. Over billions of years, many comets have changed very little, having spent most of their lives in the cold outer solar system where impacts are rare and gases do not escape. Therefore, the dirty snow of these ancient comets retains the composition of the solar system's primordial soup. These long-frozen comets give us a unique opportunity to explore a preserved sample of our own birth cloud as it was billions of years ago.

Comets contain a great deal of water as well as organic molecules needed for the development of life. Comet impacts could have brought a tremendous amount of water and the building blocks of life to the early Earth. We know that lava from primordial volcanoes carried water to Earth's surface, though probably not enough to fill the oceans. It is likely that comets made a significant contribution, perhaps 10 percent of Earth's water. This water, which began as ice in distant comets, still cycles from Earth's oceans to the atmosphere, over the land, through all life forms, and back to the ocean again—the legacy of ancient comet impacts.

Volatile materials brought to Earth by impacting comets became part of the primordial soup from which life emerged. Geologic research indicates that Earth formed at a temperature too hot for organic material to form. To survive, comet organics had to reach Earth at the right time—after the planet had cooled to the point where liquid water could remain on the surface.

Organic materials also had to survive their impact with Earth. Head-on impacts would vaporize and break down most of a comet's molecular cargo. But comets that landed in the ocean or entered Earth's atmosphere at a shallow angle could deliver some of their ice and dust to the planet.

There is another more subtle way for Earth to collect comet material. Particles shed by comets form a flattened cloud of dust and gas in the plane where the planets orbit. Sunlight reflecting off these particles creates a faint glow called *zodiacal light,* which is described in more detail in Chap. 4. In the early solar system, this comet debris cloud was much thicker. Over billions of years, tons of these particles have drifted into Earth's atmosphere and floated gently to the surface.

The molecules of life are not life itself. Most scientists agree that comets do not carry bacteria or viruses. These and other primitive life forms developed over long time periods in Earth's waters, though comets may have provided some of the organic raw materials for the development of life.

The Comet Nucleus

A comet has distinct parts: a nucleus, a coma, tails, and an ion cloud. In the 1950s, astronomer Fred Whipple described a comet's *nucleus* as a dirty snowball—a lump of ice frozen around dust and pebbles. This "dirty snowball" had never been seen directly until 1986, when the Giotto spacecraft came within 600 km of the nucleus of comet Halley. Giotto's photograph revealed the dark nucleus of comet Halley to be largely water ice combined with carbon dioxide ice, ammonia ice and methane ice. The dark black-brown dirt contains carbon and compounds of carbon and silicon. Cometary dust and ices appear to be held together loosely, like a crusty biscuit. Inside there appears to be a core of solid water ice and silicates. A crater has even been identified on Halley's black peanut-shaped nucleus, showing us that comets suffer and survive crater-causing impacts like other celestial bodies.

Furthermore, the Giotto photographs show that the comet's nucleus is made of ice erupting through a crust of dust (Plate 8 in color insert). This accumulated dust on the surface makes the nucleus very dark, dotted with bright spots which are cracks in the dirty crust where fresh ice is exposed and rapidly evaporated by the Sun's heat. About half of the nucleus is dusty, stony or metallic, while the other half is made up of ices that vaporize at Earth temperatures. Cometary nuclei are thought to be porous and to have densities much less than water; their sizes range from 1 to 50 km. Comet Halley, for instance, is 15 km long by 8 km wide by 8 km thick. The average comet has a mass less than a billionth of Earth's.

The Comet's Coma

Comets emit gases when heated in a close pass by the Sun, eventually expending their volatile gases with many passes. All comets still active have one defining characteristic: a *coma,* which appears as a soft fuzzy round glow when the comet is first detected—usually after it crosses Jupiter's orbit coming sunward (Fig. 1.3). This fuzzy ball looks like a head of hair, hence the name *coma,* which is Greek for hair. In the

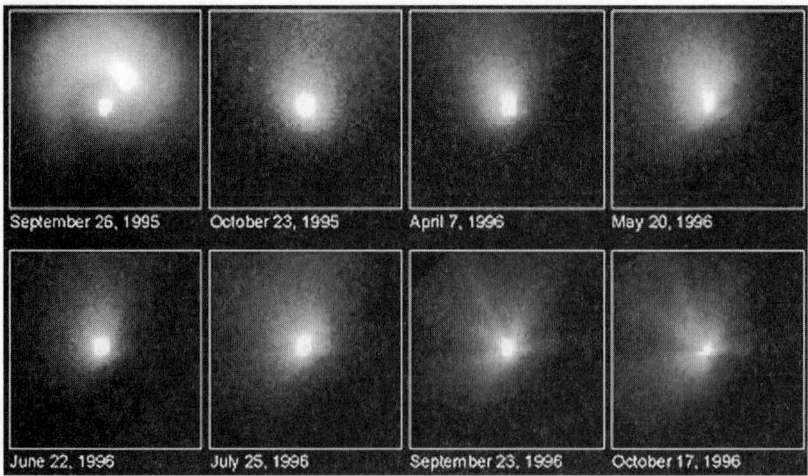

Figure 1.3 Changes in the coma and nucleus of comet Hale-Bopp. These Hubble Space Telescope observations show changes in the nucleus and growth of the coma as the comet moves closer to the Sun. The first image shows a dust outburst when the comet was beyond Jupiter's orbit. (Image courtesy of Harold Weaver, Johns Hopkins University, and NASA)

outer solar system, where temperatures are too cold for the formation of a coma, the dark, cold comet nucleus looks like an asteroid—without any noticeable activity and generally below the detectability of Earth-based telescopes.

As a comet approaches the Sun, its surface temperature increases until water ice begins to sublime—turning from ice directly to water vapor. This sublimation causes the formation of a spherical cloud of gas and dust around the nucleus. Only when this fuzzy coma appears can astronomers say for certain that a distant object is a comet. In a few comets, especially those that have not made many trips close to the sun, impurities in the water crystals may cause sublimation to occur at a much lower temperature and result in the comet's brightening when it is farther from the sun. This premature increase in brightness has caused astronomers to overestimate the size of some comets (such as comet Kohoutek in 1973; Plate 1 in color insert).

Water, ammonia, and methane stream away from the nucleus at hundreds of meters per second and cause the coma to expand to an enormous size, reaching a diameter of 10,000 to 1 million km or more. This coma is almost a vacuum by earthly standards, as evidenced by the fact that the 70,000-km thickness of comet Halley's coma was less effective at dimming the light of a background star than was a few kilometers of Earth's atmosphere.

Sunlight excites atoms and molecules in a comet's coma and causes them to glow in colors that are characteristic of different gases. We study these glowing gases by examining a comet's light with a spectrograph. The spectrograph spreads out light into a spectrum of colors, much as sunlight is refracted through a prism or raindrop. In this spectrum we find specific colors that are produced by water vapor, methane, ammonia, cyanogen, and other common molecules in the comet's coma. By studying its spectra, we can sample the contents of a comet that is millions of kilometers away.

The coma of a bright comet is surrounded by a huge cloud of hydrogen produced by the breakup of water molecules. Each water molecule contains two hydrogen atoms and one oxygen atom. When the Sun's radiation causes one hydrogen atom to leave the molecule, the other hydrogen atom is left behind with the oxygen atom. These two atoms form a smaller hydroxyl (OH) *ion cloud* closer to the coma. Skylab astronauts watching comet Kohoutek in 1973 detected water breaking down into hydrogen and the hydroxyl ion close to the nucleus and the hydroxyl ion breaking down farther out in the coma. A comet's hydrogen cloud can be greater than 10 million km in diameter.

Comet Kohoutek was also the first comet to be observed in the microwave region of the spectrum, where scientists found emissions of hydrogen cyanide and methyl cyanide. These molecules probably formed in the solar system's birth cloud and were trapped in comet ices.

The nucleus and coma combine to form the comet's *head.* The largest head probably belonged to the Great Comet of 1811, which had a coma about 2 million km in diameter. This is more than 3 times as wide as the Sun or 300 times wider than Earth. Buried deep within this coma was a tiny nucleus about 50 km wide.

Sublimation of ices also frees dust trapped in the ice of a comet's nucleus. This dust reflects sunlight and has the greatest effect on the comet's brightness. Dustier comets grow larger comas and become brighter as they near the Sun.

Nucleus Dynamics

Jets erupt in the comet's nucleus as the coma grows. These jets, spewing gas and dust into the coma, act like rockets changing the comet's speed. Forming on the side of the nucleus facing the Sun, the effect of the jets depends on how the comet's nucleus is spinning. If a comet is turning in the same direction as its orbital motion, the jets will be carried toward the trailing side of the comet. Then the jet action will cause the comet to move faster (Fig. 1.4). Imagine swinging a ball at-

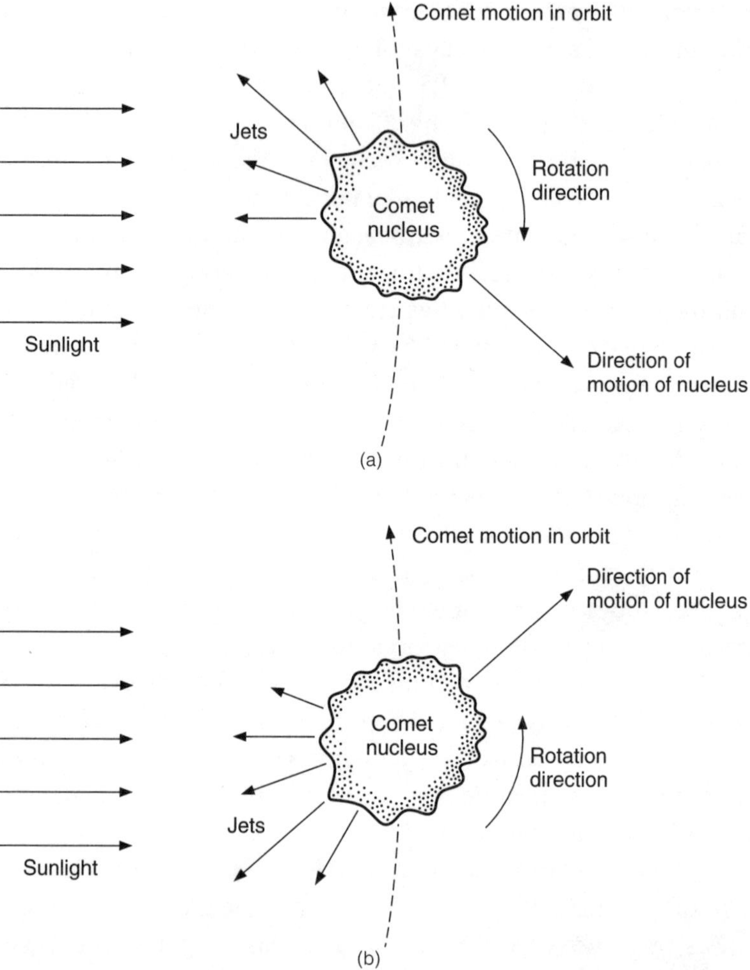

Figure 1.4 The effect of jets on a comet's motion. Eruptions of jets can cause small changes in a comet's position in its orbit. (*a*) The nucleus' direction of rotation is opposite the comet's motion in orbit, and the jets slow its orbital speed. (*b*) The direction of rotation is the same as the motion in orbit, and the jets increase the comet's orbital speed.

tached to an elastic string around your head. As the ball's speed increases, it stretches the elastic and moves farther from your hand. In like manner, the comet's increase in speed pushes its orbit outward slightly and delays its closest approach to the Sun (called *perihelion*).

In contrast, if the nucleus rotates in a direction opposite the comet's orbital motion, then the jets act to slow down the comet. The comet's orbit shrinks slightly and the comet reaches perihelion earlier. Because of its jets, comet Halley consistently returns to perihelion about 4 days later each orbit.

The Comet's Tails

As a comet crosses Earth's orbit, its tails begin to grow. Sunlight and a solar wind of charged particles push gas and dust away from the coma and cause the comet's tails to form. The outward pressure of sunlight pushing on small dust particles creates the reflective dust tail and the outward force of the solar wind produces the glowing gas tail. As active comets swing near the sun, their tails grow especially long and can extend all the way from the Sun to Earth's orbit. Most active comets develop both tails—the nearly straight gas tail and the more diffuse curving dust tail (Plate 2 in color insert).

The repulsive force of the solar wind on ionized gas is 20 to 30 times stronger than the force of gravity holding this gas to the comet. Therefore, ions stream outward so fast that the gas tail is nearly straight. The force of sunlight pushing on the dust is only 2 to 3 times stronger than gravity's pull. Therefore, dust particles move more slowly away from the coma. The orbital motion of the comet can affect the escaping dust and cause the dust tail to curve.

Ions in the gas tail glow like the gases in a fluorescent tube. Ionized carbon monoxide, for instance, causes the gas tail's prominent blue glow. This gas tail can stretch up to 100 million km into space, away from the comet's head. The dust tail shines by reflecting sunlight, so its color is distinctly yellowish. This color is most obvious when it can be contrasted with the blue tint of the gas tail.

A comet loses tons of water and other materials each day as it swings around the sun. After 100 or more orbits, the comet can no longer produce a coma or tail, but it still maintains its orbit. This inactive icy rock can easily be mistaken for an asteroid.

Where Comets Live

As early as 1644, Rene Descartes suggested that the Sun and planets coalesced from a spinning cloud of gas and dust. Descartes was correct about the formation of the Sun and solar system and his nebular theory would ultimately lead to a correct explanation for comet origins. In 1755, Immanuel Kant suggested that comets are the diffuse material at the edge of Descartes's primordial cloud, experiencing only a feeble gravitational tug from the Sun.

By Kant's time, observational astronomers had classified comets into two types, based on the time required for the comet to complete its orbit around the Sun. Long-period comets have elongated orbits that extend to the outermost region of the Sun's influence and require thousands or millions of years to complete. Short-period comets, like comet Halley, have orbits that lie within the boundaries of the planets, and their periods are up to 200 years long. All comets are accelerated by the Sun's gravitational pull and travel fastest when they are closest to the Sun. Therefore, comets spend only a few months basking in sunlight and most of their lives in the darkness of the outer solar system.

The orbits of more than 700 comets are known, and more than 100 of these have periods shorter than 200 years. These are further divided into short-period comets, with periods less than 30 years, and intermediate comets, with periods between 30 and 200 years. The orbits of these comets lie within the orbit of Pluto, with most inside the orbit of Neptune. The orbit of comet Halley extends just beyond Neptune's path and its average period is 76 years.

Perhaps a billion comets are stored just beyond Neptune's orbit in a region called the *Kuiper belt,* named after Gerard Kuiper, who postulated its existence in 1951 (Fig. 1.5). The first object in the Kuiper belt was discovered in 1992. In this region so remote from the sun, the nuclei of these comets remain frozen and most are so tiny and dark that they cannot be detected.

Pluto is the largest of these trans-Neptunian objects. Pluto's orbit brings it inside the orbit of Neptune briefly and then carries it far beyond. Pluto's orbit is tilted with respect to the plane of the solar system and is more elongated than any other planet's. In composition, Pluto is much more like an icy comet than the rocky worlds of the inner solar system or its neighboring gas giants. After Pluto's discovery in 1930, Clyde Tombaugh, its discoverer, spent decades searching for other objects at Pluto's

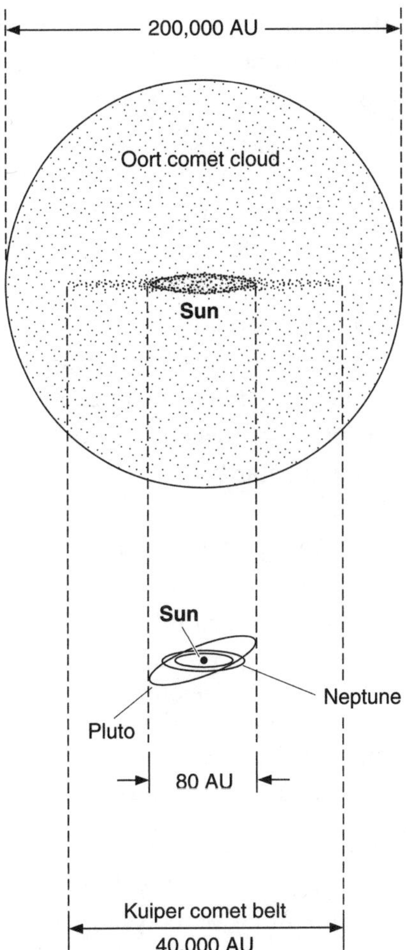

Figure 1.5 The locations of the Oort cloud and the Kuiper belt. (1 astronomical unit [AU] equals the mean distance between Earth and the Sun.)

distance and beyond and found no object at the brightness of Pluto. For 60 years, Pluto seemed to be alone with its accompanying moon, Charon. Now we have found more than 60 bodies in the trans-Neptune region and think there may be thousands more as large as comet Halley. Perhaps we should call Pluto the king of the comets.

But is Pluto also a planet? In recent years, planetary scientists have begun to describe Pluto as a *planetesimal* and no longer as a major planet. With a diameter of only 2320 km, Pluto is much smaller than all the other planets and several of the larger moons, including Earth's moon. If Pluto were discovered today, it would be given a minor-planet designation, as has happened for the trans-Neptunian objects discovered since 1992. Based on its size, Pluto would now probably be classified as a huge comet, not a major planet.

Billions of comet nuclei may also be stored in the distant Oort cloud at the edge of our solar system. The Oort cloud is more than a trillion kilometers from the Sun and extends a significant fraction of the distance to the nearest stars. Nearby stars can perturb the orbits of comets in the Oort cloud, causing some to come sunward. It is possible that the amount of dark unseen material in the Oort cloud is greater than the mass of the Earth.

The Oort cloud loses comets to nearby stars and gains comets slung outward by the gravity of the giant planets. Over billions of years, the comet birthplace between Uranus and Neptune has supplied comets for the distant Oort cloud.

When a passing star or large planet (most often Jupiter) changes the orbit of a comet, one of the following must happen:

- The comet is pushed outward from the Kuiper belt to the Oort cloud, or out of the solar system altogether.
- The comet is nudged inward from the Oort cloud or from the Kuiper belt. The comet races toward the Sun, swings around it, and goes back to the outer solar system. It may return sunward again after hundreds or thousands of years.
- The planet captures the comet and pulls it into an orbit around the planet. An example of this rare event is Jupiter's capture of comet Shoemaker-Levy 9 in 1994. Often these orbits are unstable and the comet ultimately crashes into the planet.
- The comet collides with one of the planets or moons.

Comet Cycles?

In recent years an interesting theory has surfaced concerning the mechanism for shoving comets out of the Oort cloud toward the Sun. This theory is based on the idea that comet impacts cause mass extinctions and that these are periodic. David Raup and John Sepkowski have suggested that extinctions happen every 26 to 30 million years, though the uncertainties associated with dates of extinctions are so large that this periodicity may not really exist.

Assuming that the periodicity is real, what is the mechanism for periodic comet impacts leading to mass extinctions? There are two interesting astronomical choices: the existence of another star in our solar system or the solar system's motion through the galactic plane. Perhaps an unseen distant companion star, named Nemesis, moves through the Oort cloud at its closest point to the Sun. Such a star's greatest distance from the Sun would be greater than 2 light years or almost halfway to the next nearest star. Its gravitational pull would dislodge distant comets and hurl them sunward.

Thus far we have not found the Nemesis star, and there are questions about the long-term stability of an orbit so far from the Sun and so close to other stars. But once an idea is given a name like Nemesis, it takes on a life of its own in the public mind. Tabloid publications still mention Nemesis and refer to it as the Death Star. Unfortunately, this is just too good a story to die from lack of factual support.

The solar system's motion through the galactic plane is a much more likely explanation for a 26- to 30-million-year cycle, if one actually exists. Our galaxy, the Milky Way, is flattened, with most of the stars, gas, and dust lying in its disk. We live along the inside of a spiral arm and more than halfway from the galaxy's center. The solar system circles the galaxy every 225 million years in an orbit that wobbles above and below the galactic plane. We cross this dusty plane every 30 million years. Perhaps the passage through the galactic plane produces a gravitational effect that dislodges comets and sends them sunward. Fortunately we are now halfway between passages through the galactic plane and its possible dislodging effect on comets.

Comet Viewing

New comets are discovered today at a rate of about 10 per year. Some appear on astronomical photographs made for other reasons. Others are part of an organized search program by dedicated amateur and professional astronomers. Most comets never reach naked-eye visibility. About twice a decade, a comet appears that can be seen against a dark night sky, far from city lights. Two or three comets per century will be so bright that they are visible in urban centers or in the daytime. The January Comet of 1910 and comet Ikeya-Seki, in 1965, were both visible in the daytime. Comet Hale-Bopp, in 1997, was almost as brilliant.

When a bright comet is discovered, local newspapers, radio and television news, planetarium programs, and the Internet provide viewing information. Usually viewing opportunities are in the western evening sky or eastern morning sky because a comet is brightest when nearest the Sun. If the comet is close to the horizon after sunset or before sunrise, its tail will extend upward away from the Sun and the horizon (Plate 3 in color insert).

Comet Viewing

Spectacular comets have large, faint tails which cannot compete with bright urban skies. Thus comet viewing is best on a moonless night at a place where the sky is so dark that you can see the Milky Way.

The best equipment for comet viewing is a good pair of binoculars. Two numbers are used to describe the optical characteristics of binoculars: the magnification and the width of the largest lens in millimeters. The larger this lens is, the more light the binoculars can collect and focus. This light-gathering power is much more important than the magnification—especially when viewing a large, faint comet. Assuming the optics are comparable, 7 × 50 binoculars are better for viewing comets than 10 × 35s. Low-power telescopes can also be used for comet observing, but the comet's tail may be too large for the telescope's more magnified view.

You can also become a comet hunter by joining the amateur astronomers who find and observe faint comets each year. Astronomical societies, planetariums, and observatories have information about local comet hunters. Astronomers who discover a new comet receive the honor of having the comet named after them.

Comet Photography

If a comet can be seen by the naked eye, it can be photographed using a 35-mm camera with a standard 50-mm lens, a manual exposure setting, a tripod, and a cable release. The film must be fast, at least 400 ISO. The newer 800- and 1600-speed films work very well. Print film provides more choices after taking the picture than slide film does, but the background starfield is usually much blacker on a slide.

For photography the camera should be attached to a tripod with the focus set to infinity and the lens opened up as much as possible by setting the f-stop to its lowest value (f/2.8 or lower). The exposure dial should be in the *B* or *T* position. Exposures should range from 15 seconds to 1 minute. (Longer exposures will produce star and comet trails as Earth's rotation causes the sky to appear to move.) A hood in front of the lens will help in restricting background light that can fog the film. The darkest possible background sky produces the best comet image. Dry mountaintops are better locations than humid beaches.

(continued)

Comet Memories

Everyone has a comet story—an unexpected view on an early morning, a parent dragging sleepy children outside to see something strange in the sky, a grandparent remembering a comet appearance in the darker skies of long ago. Table 1.1 contains a list and a description of the visible comets of the last 200 years, including comet Halley. If there is a personal comet in your life, this is your chance to identify it. You can also determine if a comet was in the sky during a special time in your family's history.

Table 1.1 describes when each naked-eye comet was visible and the length of its tail. To translate the tail length into something meaningful, hold out your fist at arm's length. Your fist is about 10° wide. Stack fists to imagine how long the comet's tail would look in the sky.

Table 1.1 Two Centuries of Comets

COMET	DATE OF APPEARANCE	DESCRIPTION OF TAIL
Comet Hale-Bopp	July 1996–Oct. 1997	Blue gas tail 20° long, curved yellowish dust tail 25° long
Comet Hyakutake	Early Mar.–early June, 1996	Tail at least 70° long
Comet Halley	Late fall 1985 and spring 1986	Tail 20° long
Comet West	Late Feb.–mid-Apr. 1976	Tail huge with five pieces, reddish dust tail 35° long
Comet Kohoutek	Late Nov. 1973 through Jan. 1974	Tail very faint, about 15° long
Comet White-Ortiz-Bolelli	Mid-May–June, 1970	Tail 12–15° long
Comet Bennett	Feb.–mid-May 1970	Two tails, with the longer 20°
Comet Ikeya-Seki	Early Oct.–mid-Nov. 1965	Slightly curved, brilliant, dense tail grew to 35°
Comet Seki-Lines	Late Feb.–Apr. 1962	Tail dense, bright, 15° long
Comet Wilson-Hubbard	Mid-July–Aug. 1961	Tail 25° long
Comet Mrkos	Late July through Sep. 1957	Two tails, brighter one curved and 15° long

COMET	DATE OF APPEARANCE	DESCRIPTION OF TAIL
Comet Arend-Roland	Mid-Mar.–mid-May 1957	Tail 30° long, with a sunward-pointing antitail 15° long
Eclipse Comet	Early Nov.–mid-Dec. 1948	First spotted during a solar eclipse, tail about 30° long
Southern Comet	Dec. 1947, S. Hemisphere	Tail 20–30° long
Comet de Kock-Paraskevopoulos	Mid-Jan.–Feb. 1941	Faint tail, up to 20° long
Comet Skjellerup-Maristany	Late Nov. 1927–Jan. 1928	Tail large, faint, 40° long
Comet Brooks	Late Aug.–late Nov. 1911	Straight tail, up to 30° long
Comet Beljawsky	Late Sep.–late Oct. 1911	Tail 15° long
Comet Halley	Mid-Feb.–mid-July 1910	Tail 120° long in predawn
Daylight Comet	Mid-Jan.–mid-Feb. 1910	Brilliant tail, 50° long
Great Comet	Mid-Apr.–May 1901	Two tails, a 30° straight gas tail and a 10° curved dust tail
Great Southern Comet	Jan. 1887	Tail 50° long
Great September Comet	Sep. 1882–mid-Feb. 1883	Brilliant tail, 25° long
Comet Wells	Late May–early July 1882	Tail about 40° long
Great Comet	Late May–July 1881	Tail about 20° long
Great Southern Comet	Feb. 1880	Tail about 50° long
Comet Coggia	Early June–Aug. 1874	Two tails, more than 60° long
Comet Tebbutt	Mid-May–mid-Aug. 1861	Tail as long as 120°
Great Comet	Mid-June–July 1860	Tail about 20° long
Comet Donati	Mid-Aug.–Nov. 1858	Two beautiful tails, 60° long
Great Comet	Mid-Mar.–mid-Apr. 1854	Tail about 5° long
Comet Klinkerfues	Early Aug.–early Oct. 1853	Tail about 10° long
Comet Hind	Late Feb.–Mar. 1847	Tail more than 3° long
Great March Comet	Early Feb.–early Apr. 1843	Tail 45° long

(continued)

Table 1.1 Two Centuries of Comets *(Continued)*

COMET	DATE OF APPEARANCE	DESCRIPTION OF TAIL
Comet Halley	Mid-Sep. 1835–mid-Feb. 1836	Tail about 20° long
Great Comet of 1831	Jan. 1831	Tail about 3° long when detected
Great Comet of 1830	Mid-Mar.–mid-May 1830	Tail several degrees long
Comet Pons	Late Aug.–Dec. 1825	Tail 14° long
Great Comet	July 1818	Tail 7–8° long
Great Comet	Apr. 1811–Jan. 1812	Tail 25° long
Great Comet	Early Sep.–Dec. 1807	Two tails, longer about 10°

The Cast of Comet Characters

A Comet Surprise

Comet watchers are a patient lot, happy if just one comet reaches naked-eye brightness in a year—and thrilled if a comet gets bright enough to entertain the neighborhood. Considering the growing competition of bright city lights, these awesome urban comets have become very rare.

In January 1996, a potentially spectacular comet was on its way sunward. Comet Hale-Bopp had been discovered the summer before, but would not reach its closest point to the Sun for another year. Yet comet watchers were already observing the faint comet through telescopes or powerful binoculars and watching for subtle changes.

Suddenly comet Hyakutake (named for Yuji Hyakutake) literally appeared out of nowhere. It was much smaller than Hale-Bopp, but its path would come very close to Earth. Only 5 weeks after its discovery, the comet would be within 16 million km of our planet. The announcement brought excitement and concern—the joy of exploring a new comet mixed with the realization that if this comet had been aimed directly toward Earth, we would have had only five weeks warning. With a nucleus 2 to 3 km wide (the size of a small village) and moving at 150,000 kph, comet Hyakutake would have caused major worldwide devastation if it had struck the Earth.

On March 25, the comet flew past Earth at a distance of 15.2 million km, picking up speed each day as it rushed sunward. The Sun's heat vaporized ices, which then erupted through the comet's crust. A tail of ionized gas, corralled by the Sun's magnetic field, blew straight back from the comet's head. Tail colors included the blue of carbon monoxide and a touch of red from ionized water vapor.

With a 144,000-km-wide coma and a 144-million-km tail, comet Hyakutake was typical of many bright comets orbiting the sun. But for viewers on Earth, it was spectacular, with a head four times the width of the full moon and a tail that stretched over half of the sky (Fig. 2.1).

Figure 2.1 Comet Hyakutake photographed over the desert at Villa Nueva, New Mexico. (Image courtesy of Laurel Ladwig)

Comet watchers could see gas and dust spewing out of the comet's nucleus at the rate of several tons each second. Spikes of glowing gas cast beams along the comet's tail. The comet's mushroom-shaped head plowed into the solar wind. As the comet warmed, whole pieces of its nucleus sloughed off, turned into gas, and created beadlike features that sprouted minitails of their own (Plate 4 in color insert).

As the nucleus spewed out more and more dust, the inner coma developed a yellowish hue. X-ray emissions appeared from a crescent-shaped region just in front of the comet. The Hubble Space Telescope spotted two highly energetic jets of gas and dust streaming from the nucleus.

Astronomers identified comet Hyakutake as a repeat performer. It had been in our neighborhood just 8000 years ago. A comet that returns sunward many times develops a rigid crust of debris and refrozen ice. As it approaches the Sun, the Sun's heat warms up ices below the crust and causes geysers to punch holes through the crust and form jets. First-time comets have no crust and produce a uniform outflow. This comet last appeared around 6000 B.C., but close encounters with the planets have now altered its path and the comet will wait until 18,000 A.D. to return again.

The Solar and Heliospheric Observatory (SOHO) satellite photographed comet Hyakutake as it came within 30 million km of the sun. Images showed the comet's head and three separate tails, each behaving differently as the comet swung around the Sun. A tail of heavy particles followed the comet in its orbit, intense solar radiation pushed away another tail of light dust particles, and the solar wind forced an ionized gas tail outward and lined it up with the magnetic field in the Sun's corona (Plate 5).

Back on Earth, comet Hyakutake was an awesome sight—appearing so suddenly and moving with dramatic speed across the northern sky near the familiar Big and Little Dippers. It literally sailed right over our heads—a wake-up call in our own backyard.

Comet over Bethlehem

Years ago, in response to questions from the public, chapter author Carolyn Sumners researched the history of comet appearances as a possible explanation for a bright light that appeared over Judea at the time of Christ's birth. The exact date is not certain—probably between 6 and 1 B.C. Sumners was surprised by the antiquity of the historical comet record and by the reputations that comets had already earned 2000 years ago.

Influenced by the appearance of comet Halley in 1301, the Italian artist Giotto di Bondone painted the first realistic comet as the Star of Bethlehem hovering over a stable and pointing down toward the Christ child (Plate 6). Comet Halley was in the wrong part of its orbit at the time of Christ's birth, but historical records indicate that there might have been a bright comet in 4 B.C. Comets do make great direction indicators—they can move across the sky and seem to point downward, with their tails extending up from the horizon.

Yet a comet explanation for the Christmas Star is intuitively wrong. In analyzing human interpretations of astronomical events, context is important. In ancient cultures, comets were considered harbingers of evil. Their appearances foretold war, pestilence, or the death of a ruler, not the birth of a king. Because of the bad reputations given comets, another explanation, a planet conjunction, is more likely for this famous celestial event.

This investigation brings up an interesting question: Why do comets have such a bad reputation and why did their appearance have such an effect on ancient astrologers? Was it their strangeness in an ordered starfield of objects with predictable motions? Or, as some archaeoastronomers have suggested, does this comet phobia come from a prehistoric disaster caused by the appearance and impact of a killer comet?

Cases of Comet Phobia

Perhaps ancient cultures feared comets because comets refused to play by the rules—they defied the cosmic order. Their appearances were unkempt and unexpected. Their apparent sizes in the sky dwarfed every other celestial object, and their paths were often backward or at very unconventional angles. Rarely did comets follow the stately motions of the Sun, Moon and planets through the constellations of the traditional zodiac. They appeared in the wrong place and traveled in the wrong direction at the wrong speed. Since comets couldn't be predicted, they called into question the court astrologer's knowledge and power as an expert on celestial happenings. They disturbed the order of things—

both in the heavens and on Earth. Therefore, comets became harbingers of unrest and a threat to the status quo.

Ancient civilizations kept records of these warnings in the heavens. The earliest drawings of comets are found in a Chinese silk book of 2 B.C. The earliest comet sighting on record dates to the Chinese court astrologers of 1059 B.C. when the conquering King Wu was supposedly thwarted by the appearance of a comet which aided the people of Yin and their king Chou Hsin. In fact, this may be the first recorded observation of comet Halley, although it is difficult to chart the comet's path this far backward in time.

By the fourth century B.C., Chinese astrologers had sketched 27 types of comets and described the destruction produced by each. In Chinese and Japanese records, every return of comet Halley has been identified since 87 B.C., and possibly as far back as 240 B.C., depending on how reliable our orbital analysis is. The identification of these isolated appearances every 76 years as returns of the same object had to wait until the seventeenth century.

To the ancient Greeks, comets were "hairy stars," complete with beards. Their presence in the orderly heavens was a sign of impending disaster. The Greeks had identified an order to the heavens with the planets and moon moving along the familiar zodiac path. Comets, however, rarely remained in the zodiac band.

By the fourth century B.C., Aristotle decreed that comets were atmospheric phenomena, weather events caused by vapors from Earth burning in the region of fire surrounding the planet. This interpretation did little to calm public fears.

Comets were considered particularly bad omens for rulers in power, and their appearance was taken seriously. When a bright comet appeared in 60 A.D., the Roman court astrologer encouraged young Nero to take the comet's arrival as a serious threat to his rule. Nero took no chances and massacred all the nobility who could threaten his power. Nero survived four comets in all—in 59, 60, 61, and 64 A.D. This last was comet Halley and it was interpreted as a warning of the fall of Jerusalem in 70 A.D. The Comet of 79 A.D. was blamed for the eruption of Mount Vesuvius and the destruction of Pompeii and Herculaneum.

In its closest recorded approach in 837 A.D., comet Halley came within 6 million km of Earth, with a head brighter than the planet Venus and a tail stretching over half of the sky. Four comets appeared that year, much to the dismay of astrologers around the world. The Chinese court placed its astrologers under house arrest and prohibited contact with the public. If the Emperor's power was in jeopardy from these comets, court astrologers might provide interpretations that could benefit enemies of the state.

In a comet's tail the Chinese saw the curved blade of a sword. The challenge was to determine the sword's victim—each war or revolution has losers and winners. By the first millennium A.D., Chinese astronomers had produced a record of 600 comet observations and had recognized that comet tails always point away from the Sun. They called comets *broom stars* sweeping in sudden change—often violent and foreboding, such as widespread disease, war, or political misfortune.

Comet Halley appeared in 1066, just before the Norman Conquest, and became a mixed omen—depending on whether you were Norman or Saxon. The Bayeux Tapestry is a 70-m-long woven chronicle of the Norman Conquest on which comet Halley is a prominent feature (Plate 7). All eyes stare upward and the caption reads *isti mirant stella*—"They wonder at the star." The unfortunate King Harold of England looks a bit confused and unaware of his impending doom. Comet Halley appeared in the spring of that year, well ahead of the Norman invasion of England and Harold's death in the Battle of Hastings.

In 1456, the astronomer Paolo Toscanelli plotted the motions of comet Halley and compared its head to an ox and its tail to a peacock stretching over a third of the sky. Europeans were understandably distressed because the Turks were moving westward into Europe after the fall of Constantinople. A world away, the Inca of Peru feared comets as evidence of the wrath of the sun god Inti. Comet Halley appeared in 1531, shortly before Pizarro conquered the Inca Empire.

In 1577, the Danish astronomer Tycho Brahe offered the first real challenge to Aristotle's atmospheric explanation of comets. Tycho carefully measured the changing position of a particularly bright comet that was visible over northern Europe for three months. If this comet belonged to the Earth's atmosphere, its appearance would change when viewed by astronomers around the continent. By correlating his position measurements for the comet with similar measurements taken by other astronomers, Tycho realized that there was very little difference—less difference, in fact, than that seen in different moon observations. He concluded correctly that comets must be at least four times farther away than the Moon, and definitely are not a part of Earth's atmosphere.

Those accepting Tycho's discovery still had problems plotting a comet's actual path. Tycho's determination to preserve his view of an Earth-centered solar system resulted in special spheres for the comets—a very difficult design. Tycho's contemporary Johannes Kepler believed in a Sun-centered solar system, but could not fit comet motions into this scheme either. He did not have a geometric shape for comet paths. Johann Hevelius published his version of the solution in 1668. His comets were ejected by the planets and followed curving complex paths toward the Sun that were quite elegant and also wrong.

Meanwhile, one gloom and doom prophet documented his comet worries. Appearances by bright comets in 1680 and comet Halley in 1682 led to a discourse on comet terrors called *Cometomania.* The author explained that comets inflame the air and must therefore cause barren soil and famine. The lack of food can then be blamed for increased disease and death. The author further suggested that the nobility, with their refined tastes and weakened bodies, might be particularly susceptible to the hardships resulting from a comet's appearance.

Shakespeare reflected the public sentiment that comets bring bad news for royalty:

When beggars die, there are no comets seen;
The heavens themselves blaze forth the death of princes.

—*Julius Caesar*
Act II, Scene 1

The Predictable Comets

This mixed environment of public superstition and scientific discovery provided a backdrop for the efforts of Edmund Halley. In 1682, Halley observed the comet that would soon bear his name. He realized that comet theories of the time could not explain the motions he saw. In frustration, he contacted the reclusive Isaac Newton to learn more about a possible law of gravity that could explain comet behaviors.

In 1684, Halley first met the older and more sullen Newton. Newton assured Halley that he had solved the equations for universal gravitation, but had mislaid his calculations. At Halley's insistence (and his offer to pay publication costs), Newton replicated his calculations and completed his *Principia* in 1687.

Newton had watched the comet of 1680 and saw it appear in the eastern sky as it moved toward the Sun and in the western sky as it pulled away. The comet had changed direction as it swung around the Sun—a behavior he explained with a very elongated elliptical orbit. Newton showed that comet paths were variations on the elliptical orbits of the planets first described by Johannes Kepler. At this point, it should be noted that Newton's correct assumptions about a comet's path were mixed with a very unusual interpretation of their function as divine shipments of fuel and water, sent to keep the Sun burning and to save the Earth from drying up.

Halley searched historical records for a comet whose path would bring it back within a reasonable period of time. He fussed with spotty historical accounts of comet observations for many years, looking for comet appearances at equal intervals. By 1695, Halley was convinced that the comet he had seen in 1682 had a path that matched the comets seen in 1531 and 1607. This comet was distinctive in that it went clockwise around the Sun as viewed from above the solar system (Fig. 2.2). The planets all follow counterclockwise orbits. A fourth comet appearance in 1456 matched the orbital characteristics of these earlier comets and fit into their pattern of appearing every 75 to 76 years.

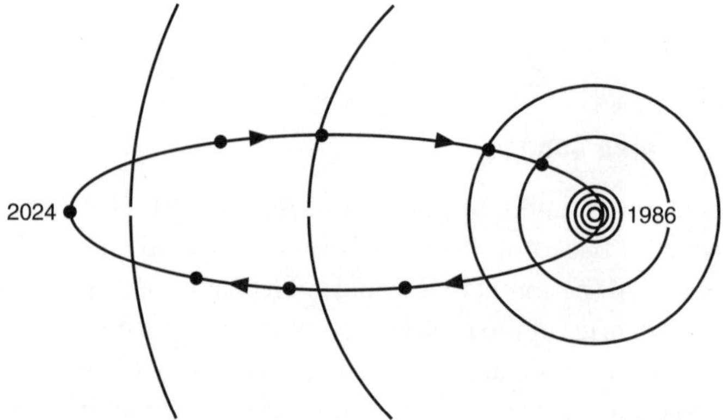

Figure 2.2 The orbit of comet Halley extends beyond the orbit of Neptune.

With confidence Halley predicted the return of this comet toward the end of 1758, a forecast which lived on after Halley's death in 1742. With all of Europe watching, the comet appeared on Christmas Day, 1758, and was named Halley's comet for the astronomer who plotted its orbit and predicted its return (Fig. 2.3).

Astronomers looked for other comets that would return on schedule like comet Halley. Johann Franz Encke tackled the observations of a faint comet seen in 1818. This comet did not follow the standard elongated elliptical path of a comet from the outer solar system. Fortunately, the field of celestial mechanics had advanced to the point that Encke could use techniques applied to asteroids to determine the comet's orbit. Without a computer the calculations were tedious and difficult, but Encke finally found an orbit that fit the observations. He determined that this comet belongs to the inner solar system and returns sunward every 3.3 years. Encke searched the records of comet

Figure 2.3 When comet Halley returned in 1986, the most spectacular view was from the Southern Hemisphere. (Image courtesy of Carolyn Sumners and Gary Young)

sightings and found observations of this comet in 1786, 1795, and 1805. He then predicted that the comet would return in 1822. The comet appeared on schedule and has been observed every 3.3 years since—except for 1944, when astronomers were too preoccupied by World War II to notice it. Comet Encke has the shortest period of any known comet.

More Comet Tales

Edmund Halley assumed that Newton's law of gravity would take the superstition and fear out of comet watching, but the comet tales continued. Napoleon Bonaparte considered comets to be mixed omens. He was born in 1769 when a great comet was in the sky. Comets appeared for several of his victories. Supposedly the Great Comet of 1811 was visible when Bonaparte decided to attack Russia. He wished upon the comet, attacked, and lost one of the biggest battles of his career.

This Great Comet of 1811 may have been the most spectacular comet to visit in historic times. It created quite a stir across the Atlantic because of a coincidence. In December 1811, a series of earthquakes rocked an area greater than 750,000 square kilometers in eastern North America, with shocks felt from New York to Florida. The great New Madrid, Missouri, earthquake followed on the heels of the Great Comet of 1811. There is no scientific connection between the comet and the quake, but the comet still got credit for the catastrophe.

By 1835, comet Halley had a reputation to uphold. It was blamed for the destruction of more than 500 buildings in New York City by fire, the massacre at the Alamo by General Santa Ana, the Zulu massacre at Weened, and wars in Cuba, Mexico, Central America, Peru, Argentina, and Bolivia. Mark Twain was born when comet Halley was visible in 1835. He often claimed that he came in with the comet and would go out with it as well. Mark Twain died when comet Halley returned in 1910.

Comet Halley reappeared on schedule in 1909, with best viewing from Earth in the spring of 1910. Between the first observation and the time it reached naked-eye brightness, another brilliant comet appeared. The Great January Comet of 1910 could be seen in the daytime and was far more spectacular than comet Halley would be. Many people confused it with the anticipated comet Halley. As it faded from view, comet Halley reached naked-eye brightness. It was much fainter, but very large in the sky because the Earth passed within 400,000 km of its tail. The public was quite apprehensive about being this close to a comet's tail—especially since poisonous cyanogen gas had recently been detected in the tail of comet Morehouse.

Certainly, the willingness to participate in mass comet hysteria is still with us and usually depends on the amount of warning we have before a comet encounter. Comet Kohoutek arrived in 1973 with a year of advance notice, along with predictions that it could be the "comet of the century." Astronomers warned that it was a first-time comet and might not be as spectacular as its first observations had indicated. But the fervor continued, coupled with predictions of the world's end. Meanwhile the comet fizzled, scarcely reaching visual magnitude. Only the Skylab astronauts got a good view (Fig. 2.4).

In contrast to a frequent visitor like comet Halley, comet Kohoutek was a virgin comet that had probably never been sunward before. At Jupiter's distance, it expelled some of its primordial material and glowed brightly. But as it neared the Sun, its frozen nucleus, long in the deep freeze of outer space, warmed very slowly and never reached the expected brightness. Kohoutek taught astronomers and the media to be wary of comet hype. As a result, the beautiful comet West got almost no attention when it dominated the predawn sky in March 1976.

The return of comet Halley in 1986 had plenty of advance billing. This date had appeared in most general science textbooks and those

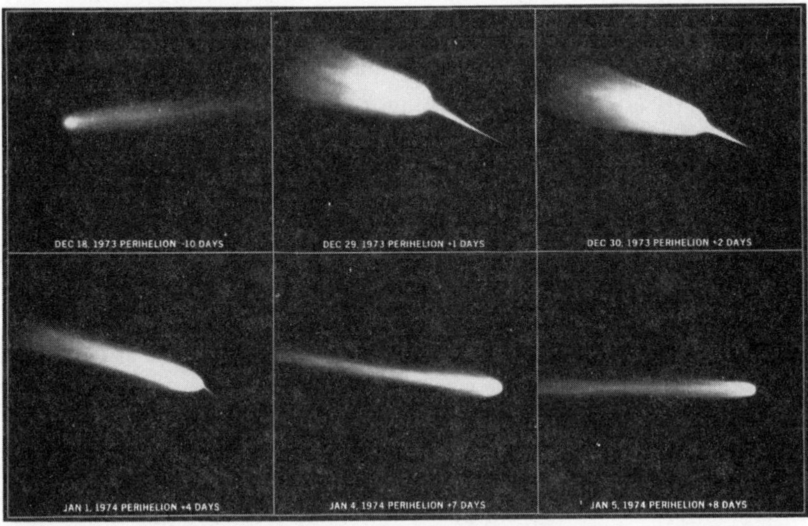

Figure 2.4 Artist's illustrations of how the orbiting Skylab astronauts saw comet Kohoutek. (Image courtesy of NASA)

who had seen comet Halley or the Great January Comet in 1910 were prepared for a repeat of the awesome show. Unfortunately, the night sky had changed in 75 years. For urban stargazers who could no longer see the Milky Way, the comet's tail could not compete with the brightness of city lights and smog. No one would see this return of the famous comet unless their sky was very dark. This time comet Halley was not favorably placed, crossing Earth's orbit when the Earth was far away. Also, the few weeks of best viewing occurred in the spring, when the comet was in the southern sky—below the horizon for most observers in the Northern Hemisphere. Many of those who did see the comet were unimpressed by the faint smudge revealed by their binoculars. In the very dark skies of the Southern Hemisphere, however, the comet did outshine the Milky Way (Fig. 2.3).

Comet watchers had to wait another decade for a brilliant comet—in this case, two: comet Hyakutake, described earlier, and comet Hale-Bopp. Comet Hale-Bopp was discovered almost two years before its close approach in the spring of 1997. Once again, years of lead time were a boon to scientists and prophets of doom alike. In the fall of 1996, a Houston amateur astronomer photographed comet Hale-Bopp near a star which he didn't recognize. Thinking that it might be an alien space ship, he called a local radio station, which eagerly reported his spaceship theory. The story spread like wildfire on the Internet. No amount of retraction could undo this popular, but erroneous, UFO sighting. The following March, 39 members of the Heaven's Gate cult in San Diego committed suicide to join the aliens they believed were behind comet Hale-Bopp. They thought that the comet was a sign to leave Earth and that the UFO was the vehicle their spirits would ride.

Like many comets before, this comet's appearance became a self-fulfilling prophecy of doom for true believers. The transformation of comet phobia into comet mania has little to do with the comet and everything to do with humanity's gift for mass hysteria. Perhaps we are blessed in some ways by the city lights and smog that hide most modern comet apparitions.

Comet Archaeoastronomy

Another strain of comet phobia has appeared in the last four centuries. In 1696, the Reverend William Whiston, a contemporary of Edmund Halley, published a treatise hypothesizing that catastrophes such as the Biblical flood occurred as a result of a comet striking the Earth. In 1950, a psychiatrist named Immanuel Velikovsky attempted to explain other Biblical events, like the parting of the Red Sea, by the close approach of an enormous comet that erupted from a volcano on Jupiter and finally became the planet Venus. Since Jupiter does not have volcanoes and Venus is far too massive to be bounced around the solar system, it is impossible to discuss Velikovsky's ideas scientifically. The law of gravity prevents real planets from behaving as he describes.

Meanwhile, a few astronomical historians have taken events like the Biblical flood seriously and have suggested that ancient historical records may contain a grain of truth. Perhaps a prehistoric comet did bequeath a curse on all future comets by crashing into the Earth and causing a global catastrophe. These archaeoastronomers cite myths and legends from cultures around the world which describe a time of flood and famine associated with fires in the sky. They hypothesize that there could have been a time when comets appeared more frequently and impacted the Earth more regularly.

Four archaeoastronomers—David Asher, Bill Napier, Victor Clube, and Duncan Steel—have associated myths of destruction from the sky with the breakup of a giant comet thousands of years ago—perhaps the parent of comet Encke. They contend that the impacts of comet fragments are the basis for the mythology associated with sky gods and for themes of generational conflict among gods.

Perhaps a word of caution is in order. Explaining ancient myths through modern perspectives is always risky business, because language syntax and cultural references provide significant latitude in interpretation. A coincidence in the historical record is interesting, but not proof of a celestial impact. These ideas remain appropriate historical speculations that lack the objective evidence to become astronomical theory.

What's in a Name?

Astronomers are faced with the challenge of finding names to match all the craters, mountains, and valleys in the solar system, as well as new moons, distant stars, and now new planets orbiting distant stars. Virtually all famous scientists, writers, musicians, artists, and a few cartoon characters have their names attached to at least one crater or rock on at least one world in our solar system. Meanwhile, nonaffiliated organizations like the International Star Registry allow you to name a faint star for yourself or a loved one, with the name inscribed in a book dutifully recorded in the Library of Congress—all for a fee, of course. There is no relationship between these names and the actual names that stars have or will have in the future, but such organizations continue to thrive. The lure of attaching your name to a heavenly object is most appealing, regardless of authenticity issues.

In a legitimate attempt to engage the public, NASA's Stardust mission to comet Wild 2 carries microchips holding the names of over 350,000 people and perhaps a few pets—all entered from the Internet. For humans, a name is very significant.

Comets offer instant fame for their discoverers. Actually there are two ways to get your name attached to a comet: You can study the appearances of many comets and discover that a single object is responsible for them all, then the comet bears your name—like Halley's comet or Encke's comet; otherwise it is a race to observe and identify a new comet before anyone else and submit your observation to the Smithsonian Astrophysical Observatory. The rules allow a comet to be named for as many as three discoverers—resulting in such names as comet Howard-Koomen-Michels (which hit the Sun, so we won't have to remember this name for the comet's return).

The lure of comet fame draws amateurs with wide-field telescopes out on moonless nights to scan the familiar starfield for an unfamiliar smudge. At the Paris Observatory, the famous comet hunter Charles Messier watched for the return of comet Halley in 1758 and went on to discover at least 12 new comets. He created the famous Messier catalog of deep-space objects as a reference for fuzzy spots that were not comets. In a dozen years Jean-Louis Pons, a porter at the Marseilles Observatory, found a dozen comets and was promoted to assistant astronomer.

Comet hunting became more lucrative in 1881, when a businessman offered a cash prize of $200 to anyone who found a comet in North America. Edward Emerson Barnard, a photographer, earned enough in comet prizes to pay off the mortgage on his cottage and moved into the ranks of professional astronomy. Barnard made the first discovery of a comet using photography. Now many comets are photographed accidentally on film shot for other reasons.

Comet Hunting

In 1998, the Smithsonian Astrophysical Observatory announced the Edgar Wilson Award—$20,000 to be divided among all amateur astronomers who discover new comets each year. The discoverer's name must be officially assigned to a new comet to qualify for a portion of the annual award.

Although comets can be found anywhere in the sky, the brightest and most easily spotted are discovered within 90° of the Sun. This zone lies in the western sky after sunset and the eastern sky before sunrise.

Observations of potential new comets should include two different sightings to confirm that the object is real and has moved through the starfield. Deep-space nebulas and galaxies remain in the same place among the stars hour after hour and smudges on an eyepiece usually disappear before a second observation. Astronomers who use photography to capture a new comet take two images, usually several hours apart. The comet will move against the background starfield during this time. Instructions on how to submit a comet discovery are found on the World Wide Web at http://cfawww.harvard.edu/cfa/ps/chat.html.

Comet hunting is still an honorable pursuit. The story is often told of Kaoru Ikeya, son of a failed Japanese businessman. Young Ikeya built his own telescope, and in 1963, at the age of 19, he found a comet to which he gave his name, thereby restoring his family's honor. The bright comet Ikeya-Seki of 1965 was his codiscovery as well.

Space Probes to Comet Halley

The 1986 return of comet Halley, with its well-documented orbit and long history, provided a unique opportunity for close-approach space probes. The Soviet Union sent two Vega spacecraft past Venus and on to the comet. The Japanese Suisei and Sakigake missions were that country's first deep-space craft. The European Space Agency's entry, named Giotto, was targeted for a close flyby of the comet's nucleus.

Halley's path rises up and over the Earth–Sun plane as the comet swings sunward. The flotilla of spacecraft intercepted the comet in March 1986, as it crossed Earth's orbit for the second time. The comet was then leaving the inner solar system and traveling tail-first. A comet is a difficult object to visit. Jets act like tiny rockets, spewing out gases and causing unexpected midcourse orbital changes; navigation computers for space probes must recalibrate as new jets appear and the comet's position changes.

The two Vega spacecraft deployed 1500-kg descent modules toward the surface of Venus and then headed for comet Halley. Vega 1 reached the comet first and measured a dark nucleus about 14 km long and much warmer than expected. Vega 2 was just 3 days behind its twin and reported that the comet's volume was about 10 times larger than predicted. The Japanese Sakigake mission measured the solar wind and magnetic field as it passed the comet. The twin Suisei probe detected cometary water, carbon monoxide, and carbon dioxide.

When the Giotto spacecraft entered the comet's coma, it discovered a velvety black peanut-shaped nucleus, with two bright jets spewing out gases. The unexpectedly dark, dusty nucleus had caused astronomers to underestimate the comet's temperature and size. Giotto detected fewer dust impacts than expected until a few minutes prior to its closest approach. Then the impact rate rose sharply, as the spacecraft apparently crossed into the path of one of the jets. Just 14 seconds before closest approach, a large dust particle struck Giotto and knocked it off alignment with Earth. It took approximately 30 minutes for the spacecraft to recover, point its antenna back toward Earth, and reestablish communications.

Giotto identified the comet's *bow shock,* where the coma plows into the solar wind (Fig. 2.5). Here the comet's hydrogen cloud forces the solar wind to flow around the nucleus, shaping the comet's gas tail streaming away from the Sun. Giotto measured the speed of the dust and gas ejected from the nucleus at 2800 kph—10 times greater than Earth's strongest hurricane.

Giotto's photos show a terrain that varies from smooth to jumbled—perhaps an indication that the comet formed from a collection of boulders. Its black nucleus is one of the darkest objects ever observed in the solar system. Giotto found a mountain range rising 400 meters and a valley running across the nucleus (Plate 8).

Combined volume and mass measurements showed that comet Halley is extremely porous and much less dense than ice. A fluffy blanket of black sooty carbon or tarlike material covers about 90 percent of the nucleus. This material is probably very porous—perhaps like a

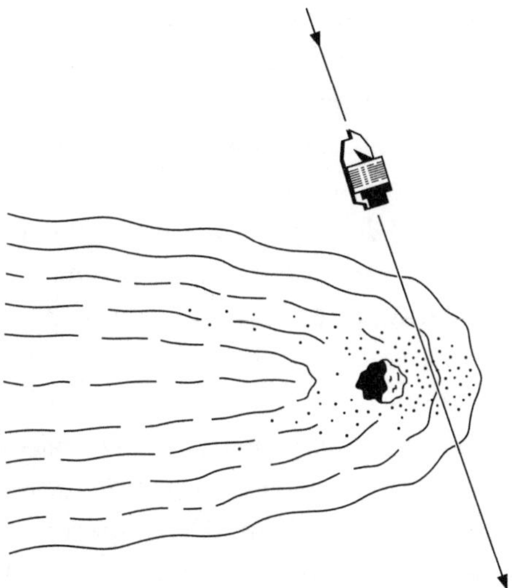

Figure 2.5 Path of the Giotto spacecraft. Giotto passed in front of comet Halley within the comet's bow shock.

black sponge or coral—and absorbs 97 percent of the light striking it. Volatile ices are trapped beneath the comet's crusty surface. Giotto sniffed the comet's jets and found large carbon-based molecules. In summary, Halley's nucleus turned out to be larger, more irregular, less dense, less homogeneous, less volatile, finer grained, and darker than anticipated (Plate 8).

Ending the Millennium with a Great Comet

The data collected about comet Halley prepared astronomers for comet Hale-Bopp. This is the comet that astronomers were watching when comet Hyakutake appeared. Now it is time to finish the Hale-Bopp story.

Astronomers had almost two years of planning time between the comet's discovery in July 1995 by Alan Hale and Tom Bopp and its close approach in March and April 1997. They could build equipment, design experiments, and schedule observing time on Earth-based telescopes and the Hubble Space Telescope. The comet cooperated by becoming bright and remaining visible for an extremely long time.

Before this return, comet Hale-Bopp's period had been 4211 years—placing it in Earth's skies in the year 2214 B.C., around the time of the great pyramids. The comet's period is now 2392 years, and Earth can expect a return visit in 4389. A close approach to Jupiter has reduced the comet's orbital period by almost half.

Estimates of comet Hale-Bopp's diameter range from 40 to 80 km, with the possibility that the comet's nucleus is elongated like a football. The Very Large Array system of radio dishes observed the nucleus at radio wavelengths and estimated its diameter at 50 km—making it 100 times larger in volume than comet Halley. At the time of discovery, the comet's nucleus was already enshrouded in a gaseous coma.

A huge cloud of hydrogen developed around the comet as it neared the Sun in the spring of 1997. The SOHO spacecraft measured the cloud's diameter at 100 million km, which is wider than the comet's visible tail. The list of ingredients for the molecular stew surrounding comet Hale-Bopp is extensive and includes neutral molecules and molecular ions. Observations of isotopes indicate that this comet is similar to comet Halley and that it was formed in the solar system.

In mid-April 1997, astronomers detected a third comet tail of neutral sodium atoms measuring more than 50 million km in length. Near perihelion, large dust grains in the comet's orbital plane produced a narrow sunward spike or antitail.

The dust production of comet Hale-Bopp was 100 times more than comet Halley's—reaching a maximum of about 400 metric tons per second. Since the nucleus is so large, comet Hale-Bopp's entire mass loss at this return was probably less than a thousandth of the comet's total mass.

Comet Hale-Bopp had everything that astronomers and the public could have wanted in a celestial show: a huge nucleus, an enormous hydrogen cloud, and three tails. Jets erupted continuously, and each image showed a slightly different comet. The observing experience night after night was spectacular (Plate 9). Only a comet close encounter could be more exciting.

Close Comet Encounters

Perils of Perihelion

The most critical close encounter for a comet is perihelion, its point of closest approach to the Sun. A comet's appearance and destiny depend on its perihelion distance and speed—the Sun's gravity can tear a comet's nucleus apart if it comes too close. The Great Comet of 1882 skimmed closer than 500,000 km from the Sun's visible surface, moving well within the solar corona. After perihelion, the comet's nucleus had split into at least four pieces arranged in a straight line—the Sun's tremendous tidal force had torn the nucleus apart.

For other comets, it is the Sun's heat at close encounter that contributes to the breakup. The short-period comet Biela, also described in Chap. 4, behaved normally until 1846, when its nucleus elongated and then split apart. On its next return in 1852, the two daughter comets appeared together, but separated by 2 million km. Neither comet has returned since. In 1976, the long-period comet West met the same fate. At perihelion this brilliant and dusty comet broke up into at least four fragments, which remained close together in one coma. In both cases, the splitting was probably caused by the comet's jets near perihelion. As these jets expelled gas and dust, the nucleus could have become very irregular and elongated until finally it split into pieces.

Heinrich Kreutz watched the breakup of the Great Comet of 1882, tracked 4 fragments, and predicted their return after 670, 770, 880, and 960 years. He also predicted that all four will plunge into the Sun at their next perihelion encounter. In 1891, Kreutz published a list of six comets which approach from behind the Sun, retreat in the same direction, and now bear his name. This century we've found even more members of the Kreutz family of sungrazers. These comets probably come from a huge parent comet that split during an encounter with the Sun many thousands of years ago. They all have perihelion distances around 500,000 km. The most famous of the Kreutz sungrazers is comet Ikeya-Seki.

Kaoru Ikeya and Tsutomu Seki independently discovered this famous sungrazing comet on September 18, 1965, within about 15 minutes of each other. Comet Ikeya-Seki soon became visible in broad daylight to anyone who blocked the Sun. Japanese astronomers on Mount Norikura saw the comet break up just 30 minutes prior to perihelion. Two definite nuclei were photographed, with a third suspected. The dust released by this comet's perilous perihelion made it a spectacular daytime object in Earth's skies.

The Solar and Heliospheric Observatory (SOHO) spacecraft has found several Kreutz comets following suicide trajectories into the Sun. On December 22, 1996, SOHO photographed the path of a Kreutz comet, curving in behind the camera's occulting mask (Plate 10 in color insert). Comet SOHO-6 failed to reappear on the far side of the Sun and must have evaporated in the Sun's atmosphere. On June 2, 1998, telescopes on board the SOHO spacecraft found two more Kreutz sungrazing comets following similar orbits. Each entered the Sun's tenuous outer atmosphere and never reappeared on the other side.

Chasing Comet Debris

All comets entering the inner solar system are changed by the Sun, although not as dramatically as a sungrazer. A comet never recaptures the material shed into its tails. If the comet has an orbit that carries this material far beyond Pluto, then the dust scatters uniformly along the comet's path and spreads throughout the solar system, with little of it reaching Earth.

The dust ejected from slower-moving short-period comets, however, remains along the comet's orbit. This debris stream is enhanced by each return of the comet. If such a comet enters the inner solar system and crosses close to Earth's orbit, there is a good chance that we will move through its debris stream once or twice each year.

The orbit of comet Halley crosses Earth's path twice—in May and again in October (Fig. 3.1). About October 21 of each year, Earth enters the inbound stream of Halley's tail particles, and on May 4 it passes through the outbound particle stream. These tiny particles become fast-moving meteors burning up in Earth's atmosphere. The famous comet Halley will not return until 2061—but you can watch pieces of the great comet burn up every spring and fall.

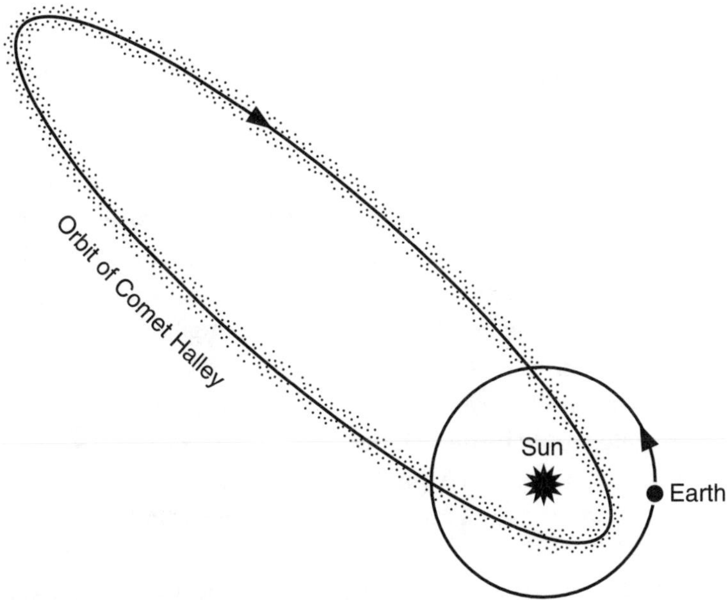

Figure 3.1 Comet Halley's orbit with meteors. Earth crosses the path of comet Halley twice each year, in May and in October. At these times, meteor debris from the comet enters Earth's atmosphere.

Earth-Crossing Threats

Comets with paths that cross Earth's orbit produce the meteor showers discussed in Chap. 4 and also pose the greatest threat to our planet. The comet producing the annual Perseid meteor shower was identified as comet Swift-Tuttle. In 1973, astronomer Brian Marsden suggested that comet Swift-Tuttle might be the same comet seen in 1737 by a Jesuit missionary, Ignatius Kegler. Marsden predicted that this comet would return at the end of 1992, and on November 7 of that year, it passed 177 million km from Earth. Armed with new observations of the comet's motion, Marsden revised his calculations and predicted the next return would occur on August 14, 2126. But if the actual date were off by 15 days from his prediction, the comet and the Earth would be in the same place at the same time.

Since comet Swift-Tuttle is about 10 km across, a possible collision should be taken seriously. Marsden continued to refine his calculations and discovered that he could trace Swift-Tuttle's orbit back almost 2000 years to match comets observed in 188 A.D. and possibly even 69 B.C. The orbit turned out to be more stable than he had originally thought, with the effects of the comet's jets less pronounced. Marsden concluded that a collision with Earth is very improbable and the comet should miss Earth by a comfortable 24 million km. However, when Marsden ran his orbital calculations farther into the future, he found that in 3044 Swift-Tuttle may pass within 1.5 million km of Earth, a true cosmic near miss.

As it crosses Earth's orbit, this huge comet is traveling at 61 kps. The explosion resulting from a collision would be about a billion times greater than the bomb that destroyed Hiroshima. A huge cloud of dust thrown into the upper atmosphere would envelop the globe for many years and cause widespread changes in climate. The effects on civilization would be devastating.

Once again, a collision is unlikely. If we take the known velocity of the comet relative to the Earth, we can determine that the comet will collide with Earth only if it is within a 3.5-minute time slot in its orbit. A difference of only 1 hour in the timing of the comet will result in it missing Earth by about 100,000 km, so a miss is much more likely than a hit.

The famous periodic comet Encke is associated with another stream of meteors called the Taurids that cross Earth's path twice a year. This stream includes 18 known asteroids as well as comet Encke.

The object which caused the Tunguska explosion in 1908 may have come from this stream. A few astronomers have suggested that the Taurid stream is clumpy, with more crowded and dangerous sections. If Earth were to drift through a busier portion of the Taurid stream, regional catastrophes from impacts might become more common.

A Real Comet Collision

The experience of colliding with a comet became much more real in the summer of 1994, as we watched a comet's suicide plunge into Jupiter. This story began in early 1993, when astronomers Gene and Carolyn Shoemaker and David Levy were conducting a photographic search for near-Earth asteroids at Mount Palomar Observatory. As described by David Levy, the night of March 23 started clear, but quickly began to cloud over. The team suspected that the night was too poor for any discoveries, but the weather had been frustrating for weeks, so they decided to photograph between the clouds. Two photographs of a region are needed to identify any faint object moving in front of the stars. The group began with an image of the sky around the planet Jupiter and then moved to other starfields. Before they could take the required second photographs, the sky began to cloud. For two cloudy hours they waited, then a hole appeared near Jupiter, and they rushed to take the necessary second photograph.

Carolyn Shoemaker analyzed the images from the weekend of photography. For region after region, the image pairs revealed only fixed stars and no moving minor bodies. At last she studied that final pair of images of the starfield around Jupiter and found an object she'd never seen before. It was definitely not a star, but more a line of lights with short fuzzy streaks stretching up from points on the line. She called it "a squashed comet". Confirming observations from other astronomers showed that this had once been a single comet which had now split into 23 little comets, complete with heads and tails. Loose dust caused a glow surrounding this comet train (Plate 11). Like normal single comets, this multiple comet received the names of its discoverers—comet Shoemaker-Levy 9 (or comet SL-9).

Astronomers began piecing together the comet's history. Comet SL-9 had been a dirty snowball in the outer solar system's Kuiper belt over 4 billion years ago. At some point, the comet was perturbed into an orbit carrying it near the Sun and crossing Jupiter's path. It probably remained in this orbit until the early twentieth century. Perhaps it was 1929 when the comet came too close to the giant planet and was captured in a very elongated orbit, swinging close to Jupiter every two Earth years. Each swing brought the captured comet nearer to Jupiter's atmosphere and made it more vulnerable to the planet's tremendous tidal forces. On July 7, 1992, comet SL-9 passed only 20,000 km from Jupiter's cloud tops. The crumbly comet came apart into little comets arranged in a row and connected by a cloud of dust. Jupiter's gravity had begun to destroy the comet. No one saw the comet break up.

Within a month of the comet's discovery, astronomers had determined its history and could make a dire prediction about its fate. On July 16 to 22, 1994, the pieces of the ill-fated comet Shoemaker-Levy 9 would crash, one by one, into Jupiter's stormy atmosphere. Observatories had just over a year to prepare for the collision—our first chance to document a comet impact. The Hubble Space Telescope with its refurbished optics would have an unobstructed view of the damage. The Galileo spacecraft on its way to Jupiter would see the actual collisions. Ground-based observatories scheduled special observing runs. Planetarium shows were written and star parties were planned. The event would be exciting and well-documented.

The size of the original comet and the composition of its fragments were undetermined. Astronomers did know that each piece would crash into Jupiter's atmosphere traveling at a speed exceeding 60 kps or 216,000 kph and would transfer a large amount of energy during impact.

Jupiter is a gas giant 11 times the size of Earth with more than 1000 times Earth's volume and over 300 times its mass. It's surface is not solid like Earth's. There may be a very compressed solid core deep inside Jupiter, but most of this giant world consists of atmosphere—growing thicker and hotter below the visible cloud tops. Astronomers wondered how much damage this string of baby comets could do to such a huge world. This would be a unique opportunity to watch the planet's thick atmosphere handle the energy delivered by each comet punch—calculated at over 25,000 megatons of TNT per fragment.

The fragments of Comet SL-9 remained intact until 10 hours before impact, when they disappeared behind Jupiter. They entered the planet's atmosphere on the night side, just before dawn (Fig. 3.2). Soon after impact, each crash site rotated into the daylight side facing Earth and the Sun.

Figure 3.2 Impacting objects. Comet fragments entered Jupiter's atmosphere one after the other on the night side of the planet. This composite was made from separate images of Jupiter and the comet. (Image courtesy of the Hubble Space Telescope Science Institute)

Each fragment was assigned a letter for the order in which it would impact the planet. Just before 4 P.M. EDT on July 16, fragment A ripped through Jupiter's clouds. As the fragment blew up, a plume rose higher than 3000 km above the cloud tops. When it fell back onto the clouds, a black bruise covered an area roughly a third the size of Earth. Fragment A was not the largest—the impacts that followed were even more spectacular.

This Hubble Space Telescope image (Plate 11) shows the train of comet fragments on May 17, 1994. Note that fragments B, F, P, and T had drifted off the straight line of the comet train. These pieces left no trace in Jupiter's atmosphere after impact. They were probably more fragile and contained more dust than the on-train fragments; the greater amount of dust made these fragments brighter than would be expected from their relative sizes. The separation of the fragments increased fivefold between 1993 and 1994, with the distance between fragments A and W growing to 1 million km. Several fragments showed signs of splitting: The P fragment was distinctly divided, and a small piece appeared beside the large G fragment. These baby comets were very fragile, yet they continuously produced enough dust to keep the spherical shapes of their comas.

Just before each fragment's impact, the surrounding cloud of dust and debris entered Jupiter's atmosphere, producing glowing flashes for up to a minute. More of the coma then pounded into the upper atmosphere in a brilliant meteor show.

The Galileo spacecraft on its way to Jupiter saw the flash of each comet fragment entering the atmosphere. The first 5-second flash came from the outer layers of the fragment beginning to burn up. Galileo also detected light from brilliant meteors reflected off trailing dust at higher altitudes.

Then the real impacts began. Each fragment entered Jupiter's atmosphere at an angle of 45° and exploded, producing plumes of hot material that rose above Jupiter's limb (Fig. 3.3). These plumes grew and expanded for several minutes.

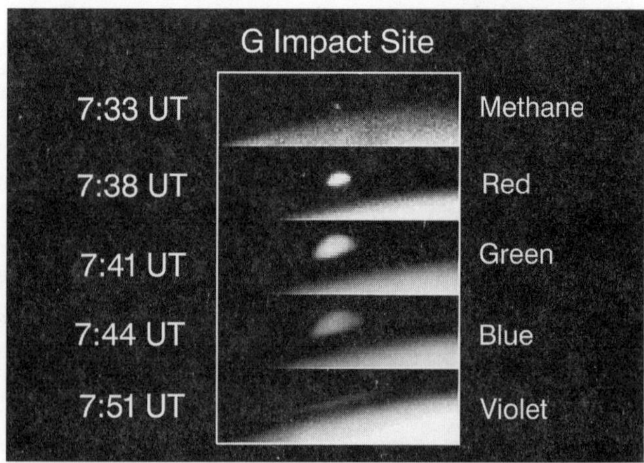

Figure 3.3 Plume from comet impact. This series of images shows a plume near the terminator at the time of the impact of the first comet fragment. The plume is 1000 to 1500 km above the limb. (Image courtesy of the Hubble Space Telescope Science Institute)

As each fragment penetrated the atmosphere, an enormous ball of superheated gas began to rise behind it. The fireball climbed upward, heated by shock energy to temperatures hotter than the Sun's visible surface. As the fireball ballooned into space and cooled, it became too faint for Galileo's cameras, but infrared telescopes on Earth continued to track its fading glow.

The plume, containing a mixture of material from the comet and from Jupiter's atmosphere, rose 3200 km above the clouds. For the G fragment, scientists estimated the fireball's diameter at 10 km and its temperature at 7500 °C. The plume fell back into the stratosphere and millions of tons of material rained down at 5 kps, heating Jupiter's upper atmosphere to 1700 °C. This infall caused a burst of heat energy. The collapsed remnants then expanded outward for about an hour, creating dark, long-lasting bruises (Fig. 3.4).

The dark centers of these spots marked chimneys through which the fireballs escaped. Debris from large fragments covered areas

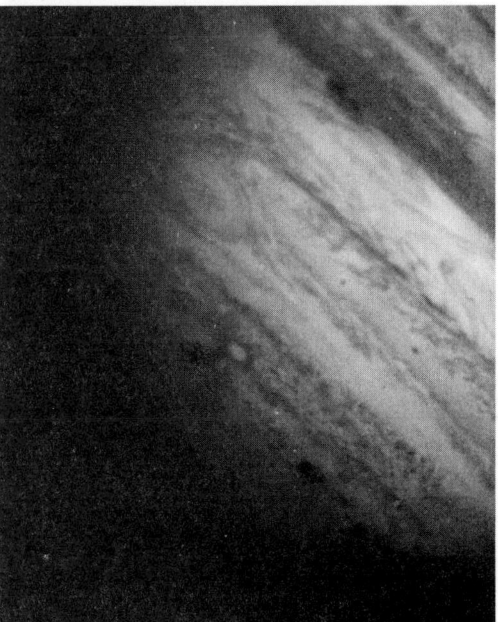

Figure 3.4 The impacts of comet Shoemaker-Levy 9. This image, taken on July 22, shows the cumulative effect of the impacts of several different fragments. (Image courtesy of the Hubble Space Telescope Science Institute)

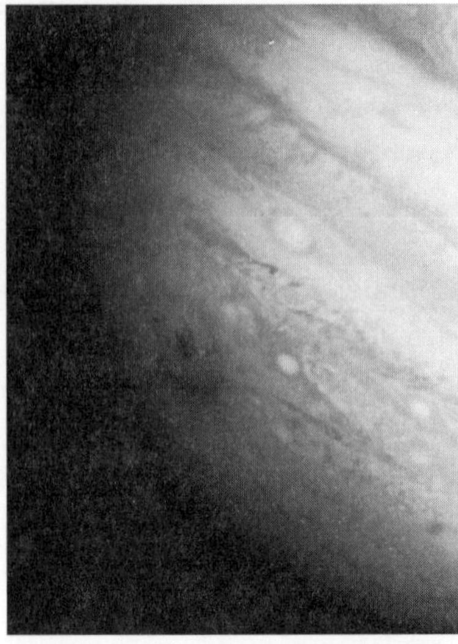

Figure 3.5 The spot and dark ring left from the impact of Fragment G on Jupiter. (Image courtesy of the Hubble Space Telescope Science Institute)

roughly the size of Earth. Amateur astronomers with small telescopes could observe each giant black eye. Spots of this size and color had not been predicted. People suddenly had ample evidence that comets do hit planets with dramatic results.

Hubble Space Telescope photos of larger SL-9 impact sites show transient rings that were probably caused by waves high in the atmosphere (Fig. 3.5). Like a powerful ocean swell, these waves caused the atmosphere to bob up and down as it rolled outward. While bobbing up, the gas cooled and dark material condensed, producing rings around the spots.

The original comet had probably been 1.5 km in diameter, with a density much less than water. Its fragments would have floated if placed gently in the middle of a terrestrial ocean. But there was nothing gentle about this impact. A 200,000-kph snowball of any size is cause for concern. The total energy released into Jupiter's atmosphere equaled about 1 million megatons of TNT.

Events such as the comet Shoemaker-Levy 9 encounter are very rare. According to Gene Shoemaker, comets which break apart before striking a planet probably come along only once every 2000 years, and Jupiter is the most likely target. Given enough time, however, Earth could also be the victim of such a strike. Suspected chains of impact craters have been identified on Jupiter's moons and on Earth and its moon.

The next impacting comet is already out there somewhere. We now recognize that the orbits of small bodies change chaotically and unpredictably. Astronomers cannot calculate their orbits forward or backward for long stretches of time, because these bodies experience close approaches to the giant planets. Small shifts caused by encounters in the outer solar system can cause dramatic changes thousands or millions of years later when an object enters the inner solar system. The long-term future of Earth is not as rigidly predictable as we thought just 25 years ago. Nature has shown us that comets do crash into planets.

Comet Probes

Recent discoveries about comets Halley, Hyakutake, and Hale-Bopp have brought many answers, followed by many more questions—puzzles that require data collected at a comet.

- How much material is inside a comet? Even when we can measure the size of a comet's nucleus, we are still not sure of the comet's composition and density.
- Is a comet a dirty snowball or an icy dirtball? Does the dark exterior cover ices mixed with dust or dust coated with ice?
- Why is the nucleus of a comet so dark? Do all comet nuclei have the brownish-black velvet hue of Halley? Is the interior as dark as the dusty surface?
- What is different about a hot spot on a comet's nucleus? Why do jets of dust and gas erupt at certain places? How do these jets cause some comets to split?
- What is left when a comet has exhausted its supply of ices and can no longer sport a coma and tail? What kind of collision threat do dead comets pose?
- How is a comet assembled—is it a solid mass or a pile of loosely joined rubble?

Obviously, there is more to come in this comet story. A host of spacecraft will soon search for these answers by visiting a variety of nearby periodic comets. These missions are described here as they are planned at the time of publication.

Deep Space 1 Reaches Comet Borrelly in 2001

Deep Space I flew close to asteroid Braille on July 28, 1999. The spacecraft is now on a trajectory that could result in a flyby of comet Borrelly in 2001. Comet Borrelly is one of the most active comets that regularly visit the inner solar system.

CONTOUR Reaches Comets Encke (2003), Schwassmann-Wachmann 3 (2006), and d'Arrest (2008)

The Comet Nucleus Tour (CONTOUR) will take images of at least three comets, make comparative spectral maps of them, and analyze the dust flowing from them. The spacecraft is scheduled for launch in July 2002, with a flyby of comet Encke at a distance of about 100 km in November 2003. Encounters with comet Schwassmann-Wachmann 3 and comet d'Arrest will follow in June 2006 and August 2008, respectively.

Stardust Reaches Comet Wild 2 in 2004

The Stardust mission (Plate 12) will return comet dust from periodic comet Wild 2 (pronounced "vilt")—a newcomer to the inner solar system. Until 1974, it orbited in frozen obscurity between Jupiter and Uranus. Then a close encounter with Jupiter perturbed its path so the comet now travels between Mars and Jupiter. By the time Stardust encounters it, Wild 2 will have made only five trips inside Jupiter's orbit. It should be in pristine condition, with most of its original dust and gas intact. Unlike the planets, most distant comets have changed very little since the formation of the solar system. But frequent comet visitors to the inner solar system have lost most of their volatile materials and no longer generate a coma or tail. Comet Wild 2 should be much more active and exciting than the typical short-period comet.

Stardust will come much closer to the nucleus of Wild 2 than Giotto's distance from comet Halley. The space probe will be able to photograph Wild 2 over a wide range of angles for three-dimensional images of the nucleus at a resolution of 30 m. With multiple filters, the cameras should be able to detect variations in color that show different minerals and surface structures.

The Stardust spacecraft will pass comet Wild 2 at a speed exceeding 22,000 kph, about 10 times faster than a bullet. At this velocity, only a *Whipple shield* (a stack of 5 sheets of carbon filament and ceramic cloth spaced 5 cm apart) can protect the spacecraft during the hours it spends in the coma. On approach, the spacecraft's camera will use a periscope to look around the shield. As the periscope's mirror becomes sandblasted, a movable mirror will swing out and supply clear images to the camera.

Catching fast-moving comet dust is no easy feat. Although the captured particles will each be smaller than a grain of sand, a high-speed capture could alter their shape and chemical composition or vaporize them entirely. To collect particles without damaging them, Stardust will use an extraordinary substance called *aerogel*—a silicon-based solid with a porous, spongelike structure in which 99 percent of the volume is empty space. When a particle hits the aerogel, it buries itself in the material, creating a carrot-shaped track up to 200 times the particle's length. Since aerogel is mostly transparent—sometimes called *solid smoke*—scientists can use these tracks to find the tiny particles. Stardust will collect less than 0.03 g of cometary dust in all, but more than 1000 of the particles collected will be large enough for scientific analysis. Of special interest are the elements associated with life (carbon, hydrogen, nitrogen, oxygen, phosphorus, and sulfur) and their compounds.

After its comet encounter, the Stardust capsule is scheduled to return to Earth in January 2006. It will hit Earth's atmosphere at 45,000 kph, traveling 70 percent faster than the reentry velocity of the Shuttle. The spacecraft is built to withstand forces up to 100 *g*. This is the first U.S. mission designed to return samples from another body since the Apollo missions to the Moon in 1969 to 1972.

The canister will be opened and stored in the planetary materials curatorial facility at Houston's Johnson Space Center—the storage location for the Apollo moon rocks and the Antarctic meteorites. Bits of comet dust will be studied in NASA's Cosmic Dust Lab and other labs around the world.

The Deep Impact Mission Hits Comet Tempel 1 in 2005

On July 4, 2005, a 500-kg copper projectile will crash into comet Tempel 1 and create a crater as big as a football field and as deep as a seven-story building. A camera and infrared spectrometer on the Deep Impact spacecraft and observers on Earth will study the resulting icy debris and pristine material from the comet's interior. The impact will occur at an approximate speed of 10 kps.

Rosetta Reaches Comet Wirtanen in 2011

In 2003, the European Space Agency (ESA) will launch the Rosetta spacecraft on a rendezvous and orbit mission to comet Wirtanen. To gain speed, the spacecraft will fly around Mars once and Earth twice and also visit two asteroids, Mimistrobell and Rodari, along the way. Eight years after liftoff, Rosetta will fly parallel to the comet's path as it approaches Mars' orbit on its way toward the Sun.

In April 2012, Rosetta will go into orbit around Wirtanen and map its entire surface as it changes during close approach to the Sun. Instruments will measure the dust and gas produced by the comet as the nucleus comes alive with jets, vents, and a halo of gas and dust.

In the comet's weak gravity field, Rosetta will perform a slow-motion dance, taking a week to circle the nucleus once. The spacecraft is designed to swoop closer to inspect possible landing sites and to drop its lander. The lander will use a drill to take core samples and hop from place to place over the nucleus. The navigation team back on Earth will run computer simulations before each new task. The team must use information gathered by the spacecraft to make decisions about how to proceed throughout the mission.

A mission as complex as Rosetta requires a predictable comet with an orbit that keeps it nearby and moving relatively slowly. Comet Wirtanen is about as predictable as comets get. It is a veteran of the inner solar system, an elderly comet no longer spewing out a torrent of dust and gas. At close approach, it will sport a modest coma and insignificant tail. Comet Wirtanen makes up for its lackluster demeanor by being convenient. Of the possible short-period comets, it offers the quickest timetable between launch of the spacecraft and completion of the mission.

Scientists have watched comet Wirtanen since 1948 and have seen encounters with Jupiter reduce the period of its orbit from 6.65 to 5.5 years. Despite many observations, we are still unsure about the comet's mass, size, and shape. It could be round or oblong, lightweight or massive, small (1 km wide) or large (10 km wide). The best estimate is around 1.5 km. With this much uncertainty built into the mission, Rosetta will earn its name. As the Rosetta stone unlocked the interpretation of Egyptian hieroglyphs, so the Rosetta spacecraft will explore the conditions of this comet and hopefully provide information to explain the behaviors of all comets and their role in the origin and history of the solar system.

Into the Future

Comets are rich resources for water and the carbon-based molecules necessary to sustain life; they could prove very useful as we leave planet Earth and begin exploring and colonizing the solar system in the twenty-first century. In addition, cometary water ice can provide hydrogen and oxygen, the two primary ingredients in rocket fuel. Comets may someday serve as fueling stations for interplanetary spacecraft.

Perhaps our ancestors were on the right track when they took comet appearances personally. Our past and future are comet-connected. Water, once carried sunward in comets, now flows through our bodies. Colliding comets may have caused the mass extinctions that removed our competition from Earth and allowed us to thrive. Now the threat of tomorrow's impact encourages us to become a spacefaring species, and the resources on comets may make our journeys possible.

PART THREE

Meteors and Meteorites

A Rain of Comet Dust

The Stars Descended Like Snow

> It would seem as if worlds upon worlds from the infinity of space were rushing like a whirlwind to our globe . . . and the stars descended like a snow fall to the earth.
>
> —Georgia *Courier,* November 1833

Comets race through space, the heat of the Sun boiling off and ionizing their ices into deep blue tails. Dust grains, trapped in those ices, become entrained in the rushing outflow of gases and form into a second tail, red to yellow in color. Thus each mote of dust, freed from its comet, becomes a microscopic body defining its own orbit around the Sun. Because each particle is so tiny, ranging in size from grains of sand down to mites smaller than specks of flour, the very pressure of sunlight itself is enough to perturb these orbits until eventually the dust spreads out, away from the comet's original path, to fill the whole solar system.

Indeed, it is possible to see this dust with the naked eye. Soon after sunset on a dark, clear moonless night—best in springtime (or just before sunrise in the fall)—look to the west (east at sunrise) for a pyramid of light no brighter than the Milky Way. This is the *zodiacal light*—sunlight scattered off tiny flakes of dust orbiting in space between the Sun and us. An even fainter manifestation, visible only on the darkest of nights around midnight, is a slight brightening of the sky in the zodiacal constellation that is directly opposite from the Sun that night. This is the *gegenschein*—light reflected from larger bits of dust stretching out to the asteroid belt.

As Earth journeys around the Sun, it is inevitable that it will encounter and sweep up these cometary orphans. The outcome of these collisions between the massive Earth and the particles of dust, at speeds of up to 90 kps, is never in doubt—the dust loses.

The last moment of the dust particle's existence is marked by a streak of light as it burns up in Earth's atmosphere. These streaks are known by many names: falling stars, shooting stars, meteors. On most clear, dark nights you can see a few meteors every hour, with occasional bright fireballs. If big enough, the sudden deceleration of a piece of comet debris as it hits the atmosphere will cause it to explode into a fireball flaming across the sky. These are often so bright that they can be seen during the day.

By itself, each meteor is a touch of beauty, an accent of light, in the black night sky (Plate 13). Occasionally, though, nature outdoes itself.

On November 12, 1799, the American natural philosopher Andrew Elliot was sailing from Philadelphia to New Orleans. At about 3 A.M., off the coast of Florida, he saw a "phenomenon grand and awful; the whole heavens appeared as if illuminated by skyrockets, which disappeared only by the light of the sun after daybreak. The meteors which appeared at any one instant as numerous as the stars, flew in all possible directions . . . and some of them descended perpendicularly over the vessel we were in, so that I was in constant expectation of their falling among us."

Though his account appeared in the *Transactions of the American Philosophical Society* in 1804, few people remembered the story until 30 years later. There had been unusually rich streams of meteors in November 1831 and again in November 1832, but nothing compared to what struck the Atlantic seaboard of North America early in the morning of November 13, 1833.

"Imagine a constant succession of fireballs, resembling rockets, radiating in all directions from a point in the heavens . . . meteors of various sizes and degrees of splendor: some mere points but others were larger and brighter than Jupiter or Venus; one was nearly as large as the moon. The flashes of light were so bright as to awaken people in their beds," wrote Denison Olmsted, a professor at Yale. Upwards of 10,000 meteors, or about 3 every second, fell between 5 and 6 A.M. that day in New England.

Perhaps most eerie to Olmsted was that "the balls, just before they disappeared, exploded . . ." and yet "no report or noise of any kind was observed."

Writing 90 years after the event, meteor science pioneer Charles P. Olivier recounted childhood stories from the family cook of the day when she was 10 years old and "saw the stars fall." He wrote, "though the shower occurred sixty or seventy years before, the impression never left her . . . and she vividly described the terror as 'the stars fell, and fell, thick as snow coming down in a snow storm. . . .' " Many who saw the shower that night believed that the Day of Judgment had surely come.

Chances of Showers

Meteor science began with that spectacular shower in 1833. For the first time, scientists took serious notice of meteors. Simply by comparing reports from many locations, Olmsted and his colleague A. C. Twining deduced that meteors were caused by objects from beyond Earth, originally in orbit around the Sun, that created their light by burning up in Earth's atmosphere.

Some of their other deductions turned out to be not so accurate, however. They badly misjudged the height and speed of the trails, placing them farther up, and slower, than modern measurements indicate. We now know that meteors generally travel anywhere from 20 to 90 kps, burning brightly at a height of about 130 km. Olmsted and Twining thought the burning occurred because the meteors were made of combustible material that caught fire in the air. They did not realize that the meteors were traveling so fast, and so deep in the atmosphere, that mere friction with the air would be enough to vaporize the dust. Indeed, the true chemical composition of this dust was still a mystery to them, and the ultimate source of the meteors remained unknown.

Whenever humans turned their gaze skyward they saw meteors. Indeed, the most primitive Stone Ager, looking up through air unpolluted by smog and city lights, certainly knew the sky better than most "sophisticated" moderns and would have seen any number of awe-inspiring meteor showers. Ancient civilizations from China to the Americas recorded the fall of meteors. These early observers knew that a meteor could come at any time; and yet, certain days of the year were more likely than others to have rich showers.

Probably the most reliable of these showers was the annual appearance of the Tears of Saint Lawrence. August 10 is the feast of this ancient Roman martyr, burned at the stake for defending the poor. (He is the patron saint of cooks and firefighters). Every year on this night, and the following four nights, sparks of light streak the sky—coming perhaps once a minute—said to represent his tears. More prosaically, modern astronomers note that these streaks generally radiate from a point in the constellation Perseus; and so, nowadays, we call the meteors of this shower the *Perseids.*

The Lost Comet Connection

Why do meteor showers occur so regularly, year after year? Why do they appear to radiate from a particular constellation? How do we know what really causes them?

Alexander von Humboldt, exploring in South America, had seen the shower of 1799 and suggested that most of the meteors flew on paths that could be traced back to one point in the sky. Other scientists ignored this suggestion of a *radiant* point until the 1833 storm. At that time, also noting this apparent origination from a single spot in the sky, Olmsted and Twining suggested that perhaps Earth was passing through a cloud of dust.

Concentrated clouds of dust don't simply hang in space, though. Each grain is subject to the same gravity forces as any planet, and so it must be moving in orbit around the Sun. What sort of orbit might this dust have?

In 1863, another Yale professor, H. A. Newton, attacked the problem from a historical point of view. He reasoned that if the Leonid

Meteor Showers

Meteor showers occur regularly throughout the year, each named for the constellation from which the meteors seem to radiate:

The *Quadrantids* come around January 3. (Quadrans Muralis is an old name for the stars between the Big Dipper and Hercules.) They provide a show almost as good as the Perseids, though shorter-lived—rarely lasting more than one night.

The *Lyrids* occur around April 21 to 22. Typically, you can expect a meteor every 4 minutes or so. They radiate away from the constellation Lyra.

From late April through early May, peaking May 4 to 6, you can look for the *Eta Aquarids.* The constellation Aquarius rises about midnight; however, these meteors tend to reach us on orbits going out, away from the Sun, and so are best seen just before sunrise.

The last week of July is the time to look for the *Delta Aquarids.* Best viewed soon after midnight, they leave long, lazy trails in the sky. This is a favorite for campers and summer vacationers. Since the shower lasts more than a week, you can usually find some time to watch it when the bright Moon is not up to blot out its streaks.

The *Perseids* in August need no further introduction. This is usually the most active shower of the year, regularly good for a meteor every minute or two, and lasting around four days (avoid moonlight). The meteors come most often after midnight.

The next good meteor shower is the *Orionids* on October 20 to 22; as name implies, the meteors appear to radiate from the constellation Orion, which rises just before midnight at this time of year. These streaks tend to be faster than the average meteor.

The *Leonids* occur nowadays around November 18. This shower was quite spectacular in 1799 and 1833—and in other years, as we shall see. Even on its off years, it is still respectable; the streaks are the fastest and brightest of any shower.

Two showers worth noting occur in December. The *Geminids,* December 13 to 15, can be as rich as the Perseids, and their trails tend to be slow and graceful across the sky. Finally, on the week before Christmas, the *Ursids* arrive—a shower of faint but graceful meteors radiating from the Dippers.

storms of 1799 and 1833 were periodic events, then history should have recorded other storms with a period of 33 or 34 years.

What is more important, though, he reasoned to a much more subtle idea. If this event was connected only with Earth, then it should occur on the same day of the year, every year. The orientation of Earth's orbit around the Sun, however, drifts slowly from year to year. The date of the meteor swarm, if it were due to bodies in orbit about the Sun, should also show a drift.

Newton's first task was to search ancient Chinese, Arabic, and European records for any unusual meteor storms in October or November. He found just such a pattern. Storms were seen on the following dates:

October 14, 931	October 24, 1533
October 14, 1002	October 27, 1602
October 16, 1101	November 8, 1698
October 18, 1202	November 11, 1799
October 22, 1366	November 12, 1833

Not only was this meteor storm a regularly repeating event, but clearly its date drifted later with time—on average, several days every century. Considering that these showers were only visible in a restricted part of the world—Europe did not see the storm of 1833—it wasn't surprising that several years in the pattern would be skipped. What seemed indisputable was that there was indeed a 33-year interval, give or take a year, between storms. In 1863, this was exciting stuff—the next storm was due in a mere 3 years. A fine shower indeed occurred in November 1866, confirming the period.

The year 1866 also saw the appearance of the next piece of the puzzle. That year, Giovanni Schiaparelli of Milan was just beginning his long and illustrious career. Lionized in his time, and world famous for his clear and provocative books on astronomy, he is most remembered today for two discoveries gone wrong. He spent years trying the very difficult observation of spots on Mercury, finally announcing that the planet spins once every 88 days; modern radar observations have shown that he was fooled, and the real period is 59 days. More notoriously, he followed up on the suggestion of a fellow Italian astronomer, Angelo Secchi, that there might be intriguing markings on Mars. In 1871 he published a book describing these *canali,* or channels. Others took this word to mean *canals,* implying vast Martian public works projects. Modern spacecraft images thoroughly discredited the romantic but incorrect idea of Martian canals.

The results of his 1866 book, *Notes and Reflections on the Astronomical Theory of Falling Stars,* have stood the test of time. In it, he took the radiants for the Perseid and Leonid showers and asked the reasonable question, if the meteors are really coming from these points in space, what does that tell us about the orientations of their orbits (Fig. 4.1). The calculations were difficult, indeed tedious. When he finally did work out the shape of the Perseid orbits, he found that these dust particles approached Earth on a path remarkably close to that taken by a well-known comet. Whatever caused the Perseids almost certainly was connected with a comet discovered in 1862, known today after its discoverers as comet Swift-Tuttle (see Table 4.1).

Emboldened by this successful calculation, Schiaparelli tried it on the data for the Leonids. Here, knowing the well-determined 33-year

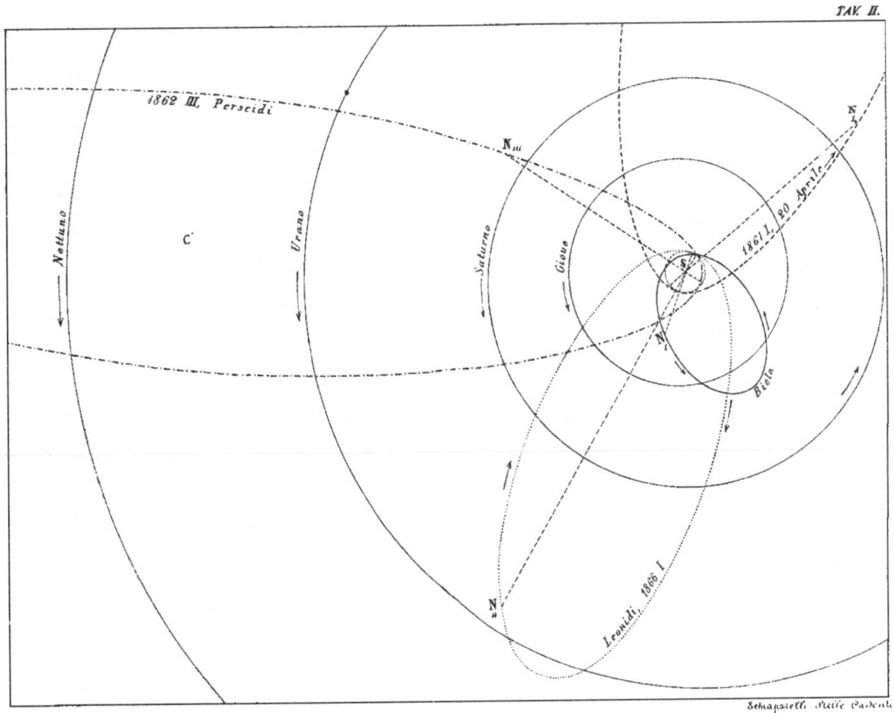

Figure 4.1 Orbits of the Perseids, the Leonids, and comet Biela. (Giovanni Schiaparelli, *Le Stella Cadenti,* 1873)

Table 4.1 Orbits of the Perseid Meteors and Comet Swift-Tuttle*

	PERSEIDS	COMET SWIFT-TUTTLE
Perihelion distance, AU	0.9643	0.9626
Perihelion longitude	343° 48′	344° 41′
Node longitude	138° 16′	137° 27′
Inclination	115° 57′	113° 34′

*What does this all mean?

Knowing the period it takes a body to complete its orbit around the Sun, you can use the laws of celestial mechanics to calculate its orbit fully. Without that period (the case for the Perseids), only some of the orbit parameters can be calculated.

Astronomers define an orbit in terms of its size, orientation, and shape.

Size: *Perihelion distance* is how close the orbit gets to the Sun. *Semimajor axis* describes the radius of an orbit in its longest direction. Both values are given in astronomical units (AU), the average distance from the Sun to Earth–around 150 million km.

Orientation: *Perihelion longitude* and *node longitude* (in degrees and minutes of arc) describe how the orbit is oriented relative to certain directions in space. *Inclination* (in degrees and minutes) describes how the orbit is tipped relative to the Earth's orbit.

Shape: *Eccentricity* defines how squished the oval shape of the orbit is; the values range from 0 (a circular orbit) to 1 (an orbit flattened into a line stretching to infinity).

period, he worked out not just the approaching path of the meteors but also the size and shape of their orbit around the Sun. Alas, no well-known comet matched the orbit he produced. His theory was cast in doubt.

Or was it? Almost simultaneously, Schiaparelli and two other astronomers, C. F. W. Peters, in Germany, and a Czech, Theodor von Oppolzer, suddenly realized that a comet newly discovered in late 1865 fit the bill exactly (see Table 4.2). The path that the Leonids followed around the Sun was, to all intents and purposes, exactly the same path followed by comet Temple-Tuttle.

The final cap to the comet-meteor connection happened hot on the heels of this discovery. This story, too, was rooted in history.

In 1772, the French astronomer Montaigne discovered a faint comet. It never grew bright enough to observe with the naked eye, so once its position and appearance in the telescope had been recorded, the comet was pretty much forgotten. His countryman Jean-Louis Pons likewise found a comet in 1805, this one much brighter. It wasn't until 1826, when the German amateur Wilhelm von Biela discovered a comet, and determined its orbit, that it became clear that all three were the same object. Comet Biela had a period of only a bit more than 6

Table 4.2 Orbits of the Leonid Meteors and Comet Temple-Tuttle*

	LEONIDS	COMET TEMPLE-TUTTLE
Period, years	33.25	33.176
Perihelion distance, AU	0.9873	0.9765
Semimajor axis, AU	10.34	10.344
Perihelion longitude	56° 26′	60° 28′
Node longitude	231° 28′	231° 26′
Inclination	162° 15′	162° 42′
Eccentricity	0.9046	0.9054

*What does this all mean? See footnote in Table 4.1.

years, orbiting between Jupiter and the Sun. More often than not it was unseen, as it traveled a path that usually took it far from Earth.

Usually, but not always. Further calculations indicated that it would pass within 45,000 kilometers of Earth on December 3, 1832. This caused a mild panic among those fearing a cosmic catastrophe but, as predicted, the comet missed Earth.

The next pass, in 1839, carried the comet on the opposite side of the Sun from Earth, where it could not be observed. The 1845 apparition was visible, however. Edward Herrick and Francis Bradley, observing at Yale, noticed something new this time. The comet now had a faint companion.

Over the next three months observers watched, fascinated, as the companion comet grew brighter, developed a tail, then two tails. The larger comet split, each piece developing its own tail, with a path of light connecting them. Clearly, something was breaking the comet apart.

The comet was next due in 1852. At that time, two comets appeared, but both were very faint. They finally grew too faint to see in September of that year. In 1859, the comet again was not observed because it was too far from Earth. In 1866, when it should have been visible, comet Biela was not to be found.

Early in December of the following year, a meteor shower called the Andromedids was particularly active. This annual shower had been known since the mid 1700s. Peaking around December 6 or 7, the swarms of meteors were exceptionally notable in 1741, then again in 1798, in 1830, and in 1838. By the 1847 shower, good enough records were kept to calculate where the meteors were coming from; these records were improved in 1867.

Note that there was no obvious 6-year periodicity to these events, and nothing to directly tie them to any comet. It was only by patiently working out an orbit for this dust, following the newly published techniques of Schiaparelli, that in 1867 both Edmund Weiss (an Austrian) and Heinrich d'Arrest (a German of French ancestry who did most of his work in Denmark) announced independently that this meteor stream had the same orbit as comet Biela.

This was startling news. Here was a meteor stream that existed even after its comet had disappeared. Schiaparelli, established as the expert in the field, had proposed that comets formed when meteor swarms coalesced into a solid object.

Instead, Weiss proposed the opposite idea. Perhaps the meteor streams arose from dust spun off of comets. The dust would follow roughly the same orbital path as the comet, but it could be easily perturbed and thus travel ahead of or behind the comet. Eventually, the comet's dust would fill the entire orbit.

Recall that in December of 1832, comet Biela had nearly hit Earth. It was clear that early December was when Earth regularly crossed Biela's orbit, whether or not the comet was there at that time. Even if the comet was not there, its dust would be. The density of dust, and so the number of meteors, should therefore be especially high near the time of the comet's passing, or once every 6 years—with enough spreading so that the year before or after the comet's passing should also provide a meteor storm visible in some part of the world. Since more often than not these storms would be visible only in remote areas, and could occur a year before or after the 6-year return of the comet, it was not surprising that no regularity to the occurrence of meteor swarms could be seen in European skies.

Weiss did one more bit of calculating. When he included the perturbing effects of the other planets on the orbiting dust, he recognized that in 1872, which by the 6-year pattern was predicted to be an especially good year for Andromedids, the planetary perturbations would move the date from early December to around November 28. He was 1 day off. On the night of November 27, 1872, observers from England to Italy counted as many as 30,000 meteors.

One of the observers in Germany, Wilhelm Klinkerfues, telegraphed to a colleague, Norman Pogson, in India: "Biela touched Earth November 27. Search near Theta Centauri." Pogson did, and discovered a faint comet. It was probably a piece of the original comet Biela, thrown off years before. It was the last piece of Biela ever observed.

The meteor storms lived on, producing memorable displays in 1885 and 1892. However, without a comet constantly feeding the orbit with dust, the intensity of the storms eventually began to wane. In 1899, observers only reported about 90 meteors per minute. Since then, the Andromedids have become only a minor shower, their 6-year storms having faded completely from sight—but not before demonstrating conclusively the connection between comets and meteor showers.

Three questions remained. First, could the other prominent showers also be connected with particular comets? The answer is yes. The Quadrantids share an orbit with comet 1491 I. The Lyrids coincide with Comet Thatcher 1861 I. Both the Eta Aquarids and the Orionids are due to comet Halley, representing the outgoing and incoming directions of the dust, respectively. The Ursids relate to comet Tuttle of 1858. The Geminids have recently been connected with the asteroid 3200 Phaeton, only discovered in 1983. It has been suggested that this asteroid is actually a comet that has completely outgassed.

Second, do the Leonid storms still return every 33 years? After 1899, the storm predicted for 1933 was pretty much a dud. However, the shower of 1966 was quite fine, with some observers seeing as many as 40 meteors per second. Between city lights and the seductions of television, though, few Americans were outdoors at the right time to appreciate it—or to be terrified by the show. After 1999, the precession of Earth's orbit and the dust's orbit are such that the Leonid storms will probably not live up to the spectacular showers of the nineteenth century.

So now we know that meteor dust comes from comets. The third question is, what is this meteor dust made of? Answering that one occupied researchers for 100 years.

Investigating Comet Dust

Many high school chemistry classes include a simple experiment: Place a chemical on a thin wire and insert it into a flame; as the chemical ignites, the students are asked to note the colors emitted. A more careful experiment, bending that light through a prism, reveals that each chemical element emits its own set of particular colors. By "fingerprinting" each chemical in this way, chemists can identify the makeup of any substance—if they can get it hot enough to glow. It would seem, therefore, that finding the chemical composition of a meteor would be a simple task. Nature does the burning for us; all we have to do is collect the colors and analyze the spectrum. Unfortunately, it is not that easy. The astronomers who tried to take these spectra—historically the Japanese, Canadians, and Czechs—ran into a number of problems.

First, you have to catch a meteor. Anyone who has gone looking for them knows that, even during a shower, it is impossible to predict just when or where the next meteor will be visible. To photograph one successfully, you have to take a lot of wide-angle exposures, over and over again, before nature obliges by putting a meteor in your field of view. In the days before electronic cameras, this meant a huge investment in film and time.

Second, you need a spectrum. The typical system is to put a large prism over the camera lens itself, turning every point of light into the colors of the rainbow. Each meteor's spectral lines appear as brighter or darker streaks crossing that band of color. Meteors come in all varieties of brightness, and likewise the spectral lines of one element from a given meteor can be thousands of times brighter than another line from a second element. Overexpose, and the brightest lines become unreadable; expose just for these lines, and you lose track of all the fainter lines. It is only recently that we have been able to computer-process some of the old photographs and extract the information of just what elements were present, and in what proportion.

A third problem is that many of the most interesting lines appear in infrared light, which ordinary film cannot detect. Indeed, trying to get a reasonable composition from a spectrum is tricky since minor constituents like sodium can produce bright, easy-to-record visible-light lines, while more abundant elements, like silicon, do not radiate as much energy in wavelengths that are easy to detect.

Finally, detailed studies of these spectra have shown that slow-moving meteors radiate their light in very different ways than fast-moving ones. The slower meteors emit a wide variety of lines, mostly yellow and red; the fast meteors appear to set up a shock wave that produces light from only a few elements, notably calcium, giving them a very blue color. In addition, the passing meteor can excite atoms in Earth's atmosphere, which then also glow. This just adds to the confusion of trying to keep track of the proportions of elements that are really present in the meteor dust.

Still, a century of meteor spectral work has yielded a general understanding of comet-dust chemistry, and a few surprises. Most of the dust seems to be made of the same elements that you would find in a typical piece of rock (or a meteorite). Some meteors, though, gave off only lines representing iron, as if they were tiny bits of metal falling at us from space. Typically, meteors reach temperatures of several thousand degrees Celsius while they glow. Most of them seem to be grains the size of beach sand.

Remote dust analysis is useful to an extent, but it is frustrating. What was really needed was a way to collect the dust before it burns up, carry it to a laboratory, and take a good close look at it. In the mid-1970s, Don Brownlee of the University of Washington figured out a way to do exactly that.

Brownlee had been studying rocks returned from the Moon by the Apollo astronauts, looking at tiny craters a few micrometers across, made by the impact of very fine dust onto the Moon. Such dust must also be hitting the Earth, he realized, but it was so small that it might actually slow down in Earth's atmosphere without heating enough to burn up. His calculations suggested that in the upper stratosphere of Earth, this tiny dust—the same material that causes the zodiacal light—could be concentrated to a million times its abundance in space. All he needed to do was find a way to fly 25 km above the surface of the Earth and collect it.

That altitude is well beyond the range of most aircraft. But in the 1950s, the military had developed a spy plane to fly over the Soviet Union at just such an altitude: the U-2. By the 1970s, satellites had mostly replaced spy planes, and a U-2 was made available to NASA for upper-atmosphere studies. Brownlee attached a small metal plate, only a few centimeters across, covered with a sticky oil to slow down and catch the micrometeorites each time this plane flew through the stratosphere. Soon, in his lab, he had a large variety of particles. Many were simply terrestrial pollution products, but some of them came from space.

Typical interplanetary dust particles, Brownlee found, come in several different varieties. Some are tiny flakes of iron; others are dry grains of well-known rocky minerals like pyroxene and olivine. The majority of the particles look somewhat like meteorite dust, except richer in light materials like carbon, nitrogen, and water (Fig. 4.2). Indeed, he found that, in detail, they are distinctly different from any known class of meteorite.

The best way to study space dust is to go out into space and collect it. Spacecraft in orbit around the Earth have regularly felt the impact of dust; indeed, in 1993 the Mir cosmonauts reported they could hear a series of impacts hitting their space station during the unusually strong Perseid shower. The Pioneer and Voyager spacecraft, the first to traverse the asteroid belt, carried simple detectors to register every time the spacecraft encountered an interplanetary dust particle. They found plenty of interplanetary dust in the asteroid belt, as well as rings of dust around all the giant planets. Detectors on the Galileo spacecraft, 20 years later, also found dust in streams radiating away from Jupiter.

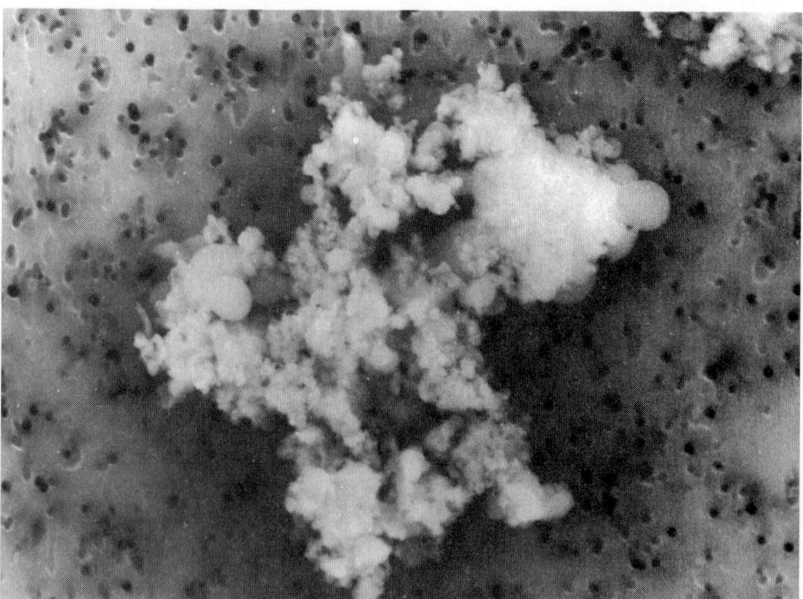

Figure 4.2 Interplanetary dust particle, 0.2 mm across, collected in the stratosphere. (Image courtesy of NASA)

An ambitious project to collect dust directly from a comet is in the works. Early in the next century, the Stardust mission (described in Chap. 3) will fly into the tail of comet Wild-2. Arriving in early 2004, it should collect at least 1000 particles fresh from the comet's tail. Then it will return to Earth, bringing this cosmic debris back to our labs in January 2006 and showing us what comet dust is really all about.

Meteorite Hunting

Rocky Pieces of History

Maybe you are stepping out your back door, crossing the street, or driving down the road when suddenly the sky brightens and you turn to see a blazing streak of light hurtling across the sky. Even though it is daylight, the fireball is visible over 100,000 square kilometers and brighter than the full moon. Within a few seconds, people in a neighboring state see the light wink out, while you, closer to the object, see a smoky trail in the sky and hear loud booms echoing through the air.

What just happened? Did a plane crash? Did you see a missile? Did a satellite reenter Earth's atmosphere?

Suddenly, before you have time to think further, you hear a whistling sound overhead just before you see a black mass crash into your garden, through the trunk of a car parked along the side of the street, or into a field next to the road.

If this happened to you, then you would be one of the lucky few to witness a meteorite fall. But what is a meteorite? What was that black mass that crashed to the ground? In most cases, you would have been watching a fragment of rock older than Earth itself—a spacefaring relic from the birth of our solar system.

Out-of-this-World Fragments of Rock and Metal

Meteorites come in all shapes and sizes. Some meteorites are tinier than dust mites, detectable only with a microscope. Because of their size, these bits of cosmic dust are hard to collect, and most scientists who study them rely on converted spy planes to sweep them out of the sky before they hit the ground. These tiny particles are dwarfed, however, by the types of meteorites typically found on the ground, which range from pebbles to objects the size of cars. The largest known specimen is the Hoba meteorite, which weighs approximately 60 metric tons. This meteorite is so heavy that it is still sitting in the ground where it fell in Namibia, Africa. Hoba may originally have been twice as heavy, but it fell so long ago that nearly half of it has disintegrated.

The largest meteorite on public display is the Cape York, from northwest Greenland. Like many meteorites, Cape York was a large iron mass that exploded into fragments when it penetrated Earth's atmosphere. Many Cape York fragments have been found, with a total weight of 58 metric tons—heavier than 140 of the Arctic polar bears that haunt that part of the world. The meteorite fell more than 1000 years ago and was first discovered by Arctic natives who used some of the iron to make harpoons, arrowheads, and knives for hunting. Europeans first learned of the Cape York meteorite when they were exploring the region in the 1800s.

In 1894, two of the Arctic natives, Tallakoteah and Kessuh, took the explorer Robert Peary to three of the iron masses. One of these weighed 31 metric tons and is the largest of the Cape York fragments. Three years later, Peary hauled the mass onto a schooner and shipped it to New York, where it now sits on display at the American Museum of Natural History.

Solar System Origins and Beyond

As impressive as these large meteorites can be, it is the fascinating geologic stories they tell that captivate scientists. These stories span the age of the solar system and, in some cases, actually push back the record of history to stars that were born long before our own Sun. These stories link the rocky remnants of our solar system's formation with astronomical observations of starbirth far away in the galaxy.

Astronomers have found newborn stars embedded in colorful masses of dust and gas called *molecular clouds.* By studying these regions of space, we have discovered that stars are produced when the dark clouds collapse into disks. As a new star grows, the nuclear furnace that makes the star shine eventually turns on, and the pressure of light pushes away much of the surrounding dust and gas. Until that happens, the birth of the star and any planets that surround it remains cloaked behind a surrounding veil of dust. What mysterious processes occur in these disks and how are they transformed into planetary systems?

Fortunately, we have the remnants of those processes in our hands. Many meteorites, older than Earth itself, were assembled from the cloud of dust and gas—the *solar nebula*—that surrounded the newborn Sun. At first the dust orbiting the young star *accreted,* or came together, to form small, loosely bound dustballs. This was normally a slow, peaceful process. However, in some regions high-temperature storms melted the dustballs, filling space with a fiery rain of molten droplets. Some scientists argue that great bolts of lightning melted the dustballs. Others contend that intense flares, similar to solar flares, were responsible. Still others argue that shock waves radiated through space, producing zones of very high temperatures. In any case, soon after the fiery rain, the molten droplets cooled and crystallized, producing millimeter-sized spheres of rock called *chondrules* (Fig. 5.1). In other areas of the solar nebula, or at other times, temperatures rose so high that some of the dust vaporized, leaving behind only small grains, the stable residue of this heating phase. Sometimes temperatures plummeted so low that rocky material condensed from the gas, much as snow condenses from cold, moist air. After a series of turbulent storms, the chondrules, stable residue, and condensates began to accrete to form small rocky bodies—the source of the meteorites that we call *chondrites.* From these materials scientists have deciphered some of the mysteries from the earliest history of our solar system.

Recently, scientists discovered that meteorites containing chondrules hold not only the first geologic products of the solar system, but also rare grains of interstellar dust. This extraordinary material did not originate in our solar system; rather, these grains were produced around stars born before the Sun. At some point they were blown into space, and later were captured in the cloud of debris that collapsed to form the solar nebula. This *stardust,* only a few micrometers in diam-

Figure 5.1 Chondrules–primitive building blocks of meteorites. (Image courtesy of NASA)

eter, includes crystalline diamond, silicon carbide, and graphite. From the chemical and isotopic properties of these grains, scientists determined that several different stars contributed material to our solar system. A violent supernova explosion that sent a shell of debris expanding through space produced most of the diamonds. Silicon carbide and some of the graphite grains appear to have come from red giant stars that were up to 3 times more massive than our Sun. The rest of the graphite came from explosive outbursts of dust and gas produced by dying stars, and from carbon-rich stars that ejected shells of gas and dust with speeds of thousands of kilometers per second.

Imagine the spectacle that accompanied the birth of our solar system. Space was filled with cold clouds of dust and gas, mixing with hotter debris ejected in glowing shells from exploding stars. While some gases were swirling in place, other gases, carrying a cargo of stardust, hurtled across space and crashed into anything in their path. On human time scales a lot of this activity would seem slow-paced, but in astronomical time scales it was a seething cauldron of icy-cold and fiery-hot material, with old stars dying and new stars forming from their remains. In the midst of all this chaos, our own local cloud began to collapse, producing a disk with an ever-growing central mass—a small protostar. The cosmic processes that led to our Sun's birth had begun.

The Growth and Destruction of Planetesimals

After our local cloud collapsed to form the solar nebula, the Sun continued to grow, blowing away or consuming most of the surrounding dust and gas. The small amount of material left behind formed a residual disk. Chondrules and other dusty components accreted into hundreds of thousands of small solid bodies. Scattered throughout the solar system, these *planetesimals* ranged in diameter from a few tens of meters to more than 1000 km. In a few specific locations, the planetesimals accreted to produce planets. However, the most massive planet, Jupiter, disrupted the accretional process in its neighborhood. Between the orbits of Jupiter and Mars, the planetesimals never accreted to a large planet. Jupiter's gravity stranded many planetesimals of distinctly different types—predecessors of the asteroids (Fig. 5.2).

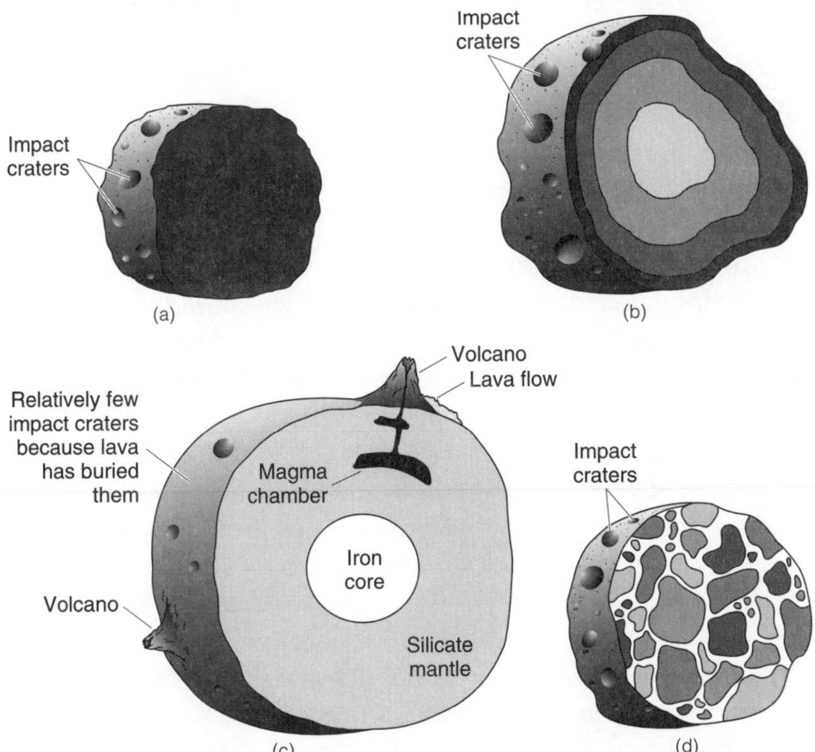

Figure 5.2 History of planetesimals in the solar system: (*a*) primitive planetesimal, (*b*) metamorphosed planetesimal, (*c*) differentiated planetesimal, and (*d*) rubble-pile asteroid.

As these stranded bodies continued to evolve, internal heat and gravity began altering the chondrule-rich rocks. Some of these planetesimals became so large that their internal temperatures rose like a giant oven, baking the rock buried beneath their surfaces. The energy needed to produce the rising temperatures came from impacts and short-lived radioactive isotopes. Dust and rock insulated the interiors of the planetesimals, much like a thick blanket. This insulation eventually led to temperatures of several hundred degrees Celsius or more.

The increased temperatures caused minerals in the rocks to recrystallize and often grow larger, producing metamorphic textures. In some cases, temperatures rose so high that the rocks began to melt, causing igneous processes such as lava flows. Some bodies grew so hot that heavy metallic elements sank to form a core and light minerals rose toward the surface. This process of *gravitational differentiation* is common in larger planetary bodies—Earth's core, mantle, and crust were formed this way.

While metamorphic and igneous processes altered many of the larger planetesimals, all of these objects were affected by collisions among themselves. Because the planetesimals were the size of mountains and, in some cases, continents, the collisions were planet-shattering events that transformed the planetesimals into many of the fragments of the asteroid belt.

Smaller impacts produced craters on the planetesimals and asteroids. Impacts also deformed, melted, and sometimes vaporized material on these surfaces, producing layers of debris covering the bodies. Such debris, called *regolith,* is the broken-up, partly shattered, partly melted layer of rubble resulting from repeated impacts over billions of years.

The Nature of Planetary Debris

Most meteorites, 94 percent of them, are classified as *stony* (Fig. 5.3). Some are fragments of asteroids almost unchanged since the birth of our solar system, while others were derived from the crusts and mantles of asteroids altered by differentiation. Stony meteorites include samples that are packed with chondrules, and show evidence of melting and impacts long ago. A small fraction of meteorites, about 5 percent, are composed of *iron* and *nickel* (Fig. 5.4). These came from the metallic cores of differentiated asteroids. Only 1 percent of the mete-

Plate 1 Comet Kohoutek. This false-color image shows luminosity differences in different parts of the comet. (Image courtesy of J. Lorre, Jet Propulsion Laboratory, and NASA)

Plate 2 Comet Hale-Bopp. Note the wide yellow dust tail and narrow blue gas tail. (Image courtesy of Barbara Wilson)

Plate 3 The tail of comet Hale-Bopp stretches upward away from the setting sun over Elm Lake in Brazos Bend State Park. (Image courtesy of Charles Gray)

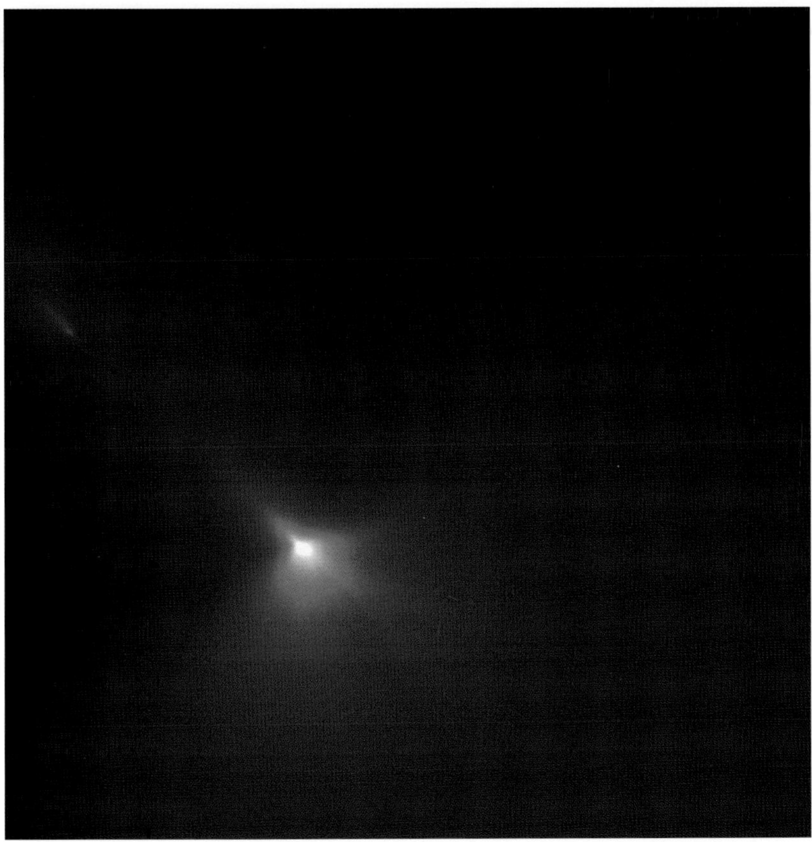

Plate 4 Nucleus of comet Hyakutake. This image shows a very small region of the comet's nucleus. At the upper left, three small pieces have broken off the comet and are forming their own tails. (Image courtesy of H. Weaver, Hubble Space Telescope Science Institute, and NASA)

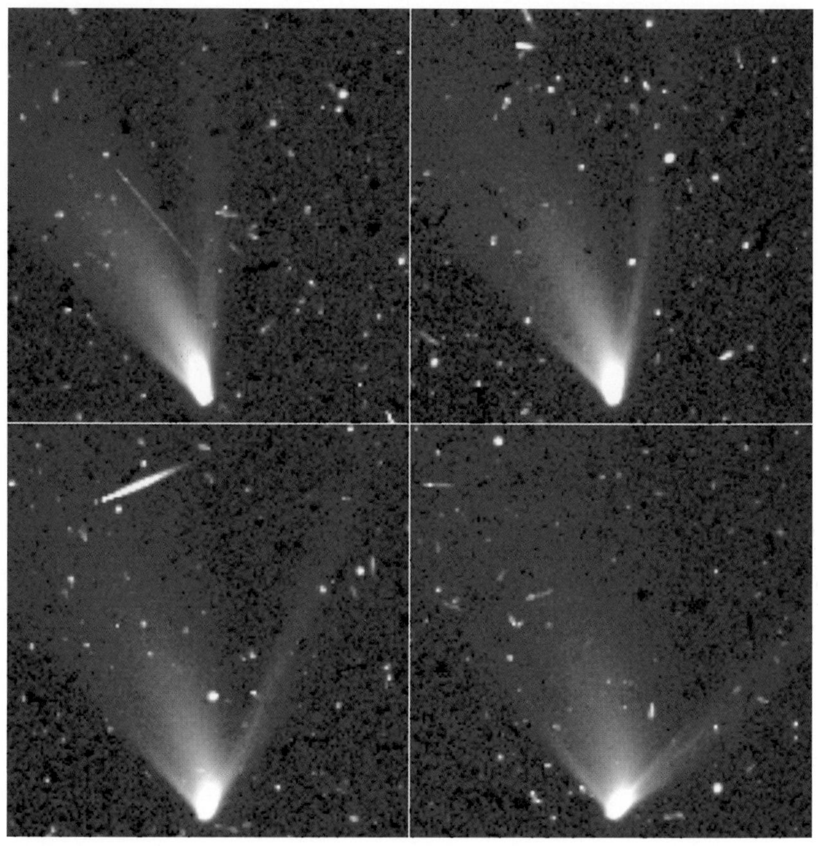

Plate 5 Comet Hyakutake at perihelion. SOHO images of Comet Hyakutake taken from solar orbit on May 2 to 5, 1996. (Image courtesy of the Solar and Heliospheric Observatory, ESA, and NASA)

Plate 6 Halley's Comet appears as the Star of Bethlehem in Giotto's *Adoration of the Magi* in the Arena Chapel at Padua. (Image courtesy of the Giraudon, Paris)

Plate 7 The Bayeux tapestry. This tapestry shows the 1066 appearance of comet Halley above King Harold. (Image courtesy of the Giraudon, Paris)

Plate 8 Sixty Giotto images were combined to produce this image of the nucleus of comet Halley. Notice the bright jets and black surface. (Courtesy of the Space Research Institute–Alan Delamere and Harold Reitsema, Ball Aerospace Systems Division, and the Max Planck Institut für Radioastronomie, Bonn)

Plate 9 Artist and author Dan Durda captured this view of comet Hale-Bopp over the Sonora Desert. (Image courtesy of Dan Durda)

Plate 10 Christmas comet. The sungrazing comet SOHO-6 approaches the Sun. (Image courtesy of the Solar and Heliospheric Observatory, ESA, and NASA)

Plate 11 The Hubble Space Telescope photographed this train of tiny comets that were once the single comet Shoemaker-Levy 9. (Image courtesy of the Hubble Space Telescope Science Institute)

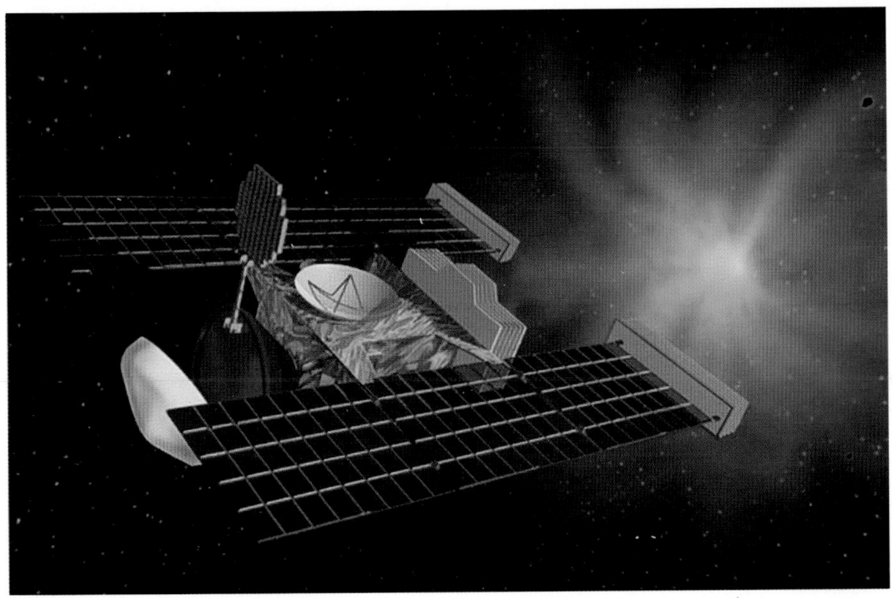

Plate 12 The Stardust spacecraft and its destination comet. (Image courtesy of the Jet Propulsion Laboratory and NASA)

Plate 13 Meteor trail on a starfield. (Image courtesy of NASA)

Plate 14 Car trunk smashed by the Peekskill, New York, meteorite. (Image courtesy of David Kring)

Plate 15 Gold Basin, Arizona, meteorite, with penknife for size comparison. (Image courtesy of David Kring)

Plate 16 Scientists discover a meteorite on the ice. (Image courtesy of NASA)

Plate 17 Nakhla meteorite thin-section photograph. (Image courtesy of David Kring)

Plate 18 Lafayette meteorite thin-section photograph. (Image courtesy of David Kring)

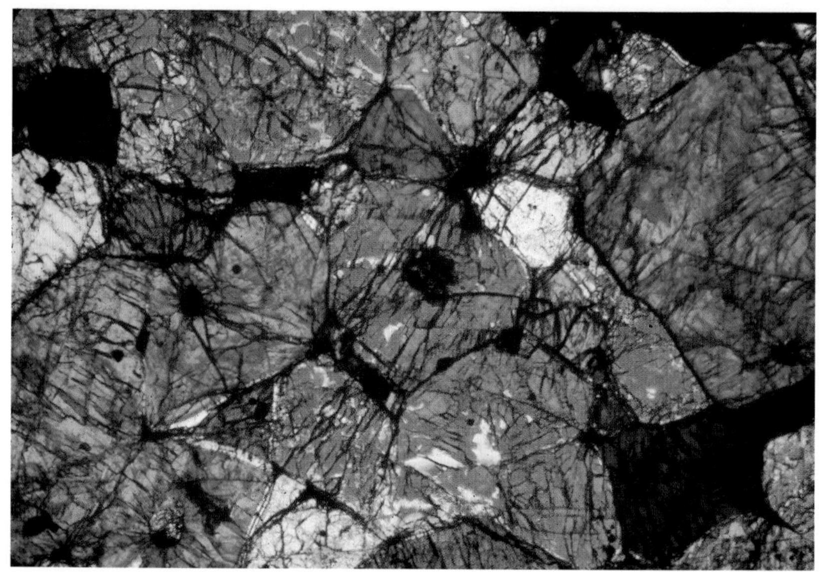

Plate 19 Chassigny meteorite thin-section photograph. (Image courtesy of David Kring)

Plate 20 LEW 88516 meteorite thin-section photograph. (Image courtesy of David Kring)

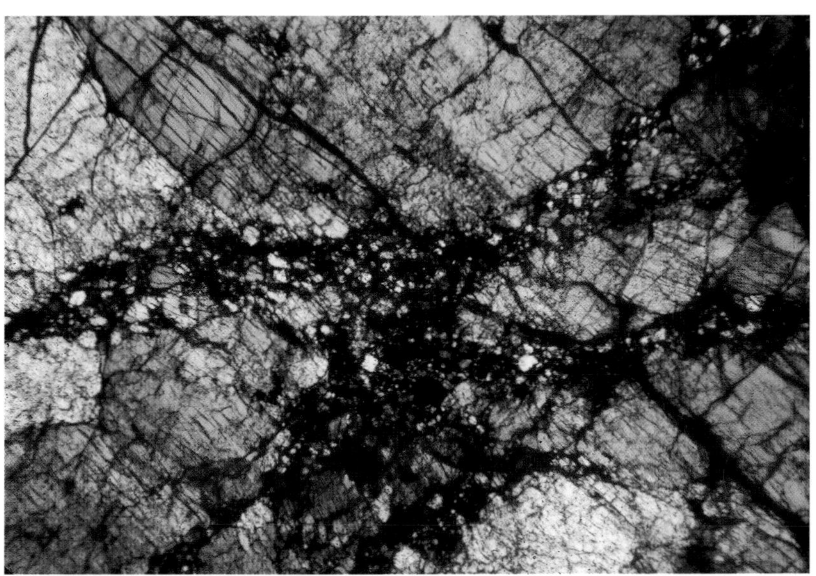

Plate 21 ALH 84001 meteorite thin-section photograph. (Image courtesy of David Kring)

Plate 22 The Martian surface; photograph by Mars Pathfinder. (Image courtesy of NASA)

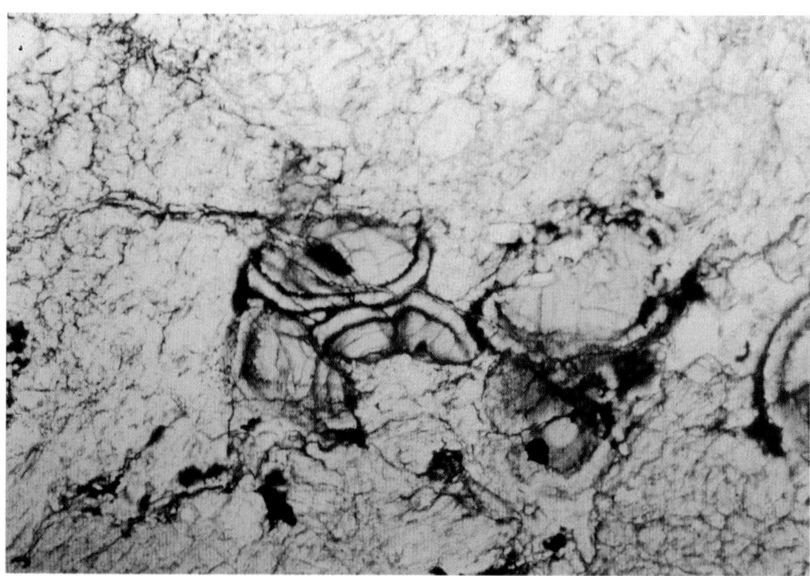

Plate 23 Carbonate globule (1 mm across) in meteorite ALH 84001. (Image courtesy of NASA)

Figure 5.3 Stony meteorite. (Image courtesy of NASA)

Figure 5.4 Iron meteorite. (Image courtesy of David Kring)

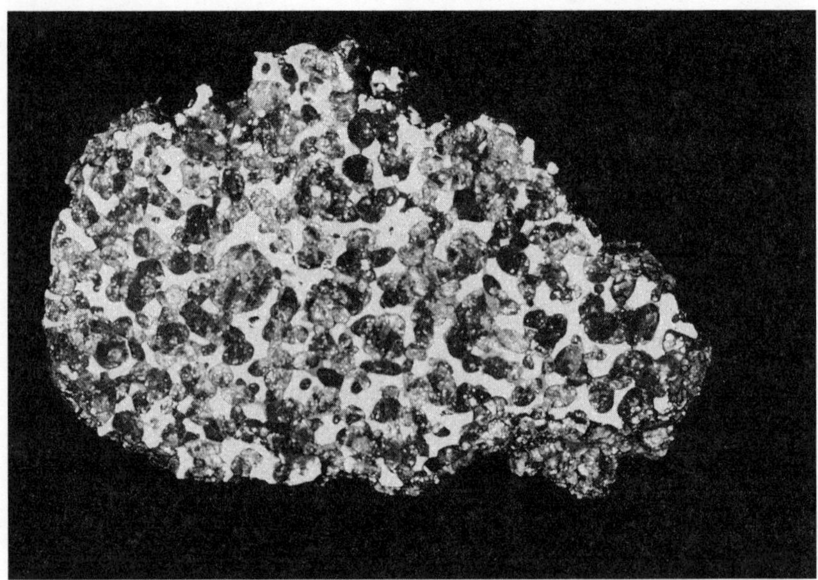

Figure 5.5 Stony iron meteorite. (Image courtesy of NASA)

orites are mixtures of rock and metal, called *stony irons* (Fig. 5.5). The stones, the irons, and the mixed meteorites give us insight into the deep interiors of larger planets, including our Earth.

It may seem inconceivable that these ancient fragments of solar-system history are found while hiking, riding a horse through the mountains, or plowing a field. They can and often are found by people going about their daily business. If you find a strange rock or chunk of iron, how would you determine if it is a meteorite?

A key piece of evidence for an object's extraterrestrial origin is the *fusion crust,* a thin black or dark brown glassy coating left after the rock's fiery transit of Earth's atmosphere. Meteorites hit the top of our atmosphere at velocities in excess of 40,000 kph. At those speeds, there is tremendous friction with atmospheric gases, which produces enough heat to melt the outermost layer of the rock. As the first layer of molten material slips off the meteorite, another layer melts, and the rock progressively erodes. In most cases, a rock completely burns up in the atmosphere or is catastrophically shattered as it rams into denser air near the ground. Sometimes, though, the object survives this fiery passage and reaches Earth's surface. Even in those cases, a lot of the original mass may have been lost. The Lost City, Oklahoma, and Innisfree, Alberta, meteorites, which fell in 1970 and 1977, respectively, lost about two-thirds of their original mass passing through the atmosphere.

The fusion crust is a distinctive feature of fresh meteorite falls and is good evidence of an object's cosmic origin. Unfortunately, there are geologic processes that can produce black layers of material around rocks, which often leads to the misidentification of Earth rocks as meteorites.

For other clues, one needs to delve inside the object. Generally, the interior of a meteorite contains lighter-colored, more crystalline material than the fusion crust. Most meteorites, even the stony variety, contain iron-rich metal. In the case of ordinary chondrites, which are the most common type of stony meteorite, these metal particles are often around a millimeter in size. They are clearly visible on broken, cut or polished surfaces of a sample. This metal is silver in color, like polished sterling silver or a chrome automobile bumper, and will attract a magnet. Iron meteorites are generally denser than Earth rocks and feel heavier than expected when you pick them up. The best way to learn about meteorites is to see them and, if possible, hold them in your hands. Most major museums have meteorite exhibits that display fusion crusts, the tell-tale metallic interiors, and other features characteristic of meteorites.

Truly Ancient Artifacts

When people find new meteorites, their first question is often, "How old is it?" It usually astounds people to learn that the answer is billions of years—often eclipsing the age of Earth.

As described earlier, the most primitive meteorites—chondrites—are composed of material produced from a spinning disk of dust and gas that orbited the newborn Sun. The first geologic process in the history of the solar system produced this material. In other words, when you read that the age of the solar system is nearly 4.6 billion years, that number was determined by measuring the ages of the chondrites. We know from studies of iron meteorites that the cores of planetesimals from the region between Mars and Jupiter froze about 4.5 billion years ago, or within 100 million years of the solar system's formation. The asteroids were geologically dead while Earth was still forming.

The age of a meteorite is the length of time since it solidified. Scientists determine this age by measuring the amount of slowly decaying radioactive material remaining in the rock. Not only can researchers determine how old a meteorite is, they can also measure how long it was in space and how long ago it landed on Earth. The travel time between its home asteroid and Earth is determined by measuring the effects of cosmic rays. This radiation, from the Sun and stars, produces changes in the nuclei of atoms near the surface of a meteorite once it has been dislodged from its parent body. Most cosmic rays cannot penetrate our atmosphere, so the radiation effects stop once the object lands on Earth. The accumulated cosmic-ray damage gives us a clock to measure the exposure age of the meteorite in space.

In addition, some of the isotopes produced by cosmic rays in space are unstable, so they begin to decay after the rock has landed on Earth. By measuring the remaining isotopes, scientists can determine how long ago the sample landed. Some meteorites are picked up right after they fall, but others landed thousands of years ago. For example, a meteorite recently found in the Gold Basin of Arizona fell nearly 20,000 years ago, during the last ice age. Many meteorites collected in Antarctica have been on Earth more than a million years.

Looking for a Piece of the Cosmos

People who want to collect meteorites are always asking scientists whether meteorites fall in particular areas of the world. For example, do more meteorites fall in North America than in Africa?

For all practical purposes, the distribution of meteorite falls is random. They land in all regions of the Earth in nearly equal abundance. Where there are more people or where people occupy a larger fraction of the land, the chances of a meteorite fall being witnessed are much greater than in less populous areas of the world. Likewise, a meteorite that falls in a city is much more likely to be observed and reported than a meteorite that falls in the middle of a desert or jungle.

Only a small fraction of the meteorites that fall to Earth are actually recovered. Many meteorites fall in the ocean, desolate places, or at times when there are few witnesses. On the basis of observations made by an automatic tracking camera system in Canada, scientists can estimate the total number of meteorites that actually fall, even though most of them are not observed by people on the ground. These studies indicate that approximately 30,000 meteorites weighing more than 0.1 kg fall on Earth each year. Of that total, only about 4400 are larger than 1 kg and only about 660 are larger than 10 kg. The threat from large meteorites is discussed in Chap. 9.

In North America, nearly 1400 meteorites larger than 0.1 kg fall each year. Of these, about 200 are larger than 1 kg and about 30 are larger than 10 kg. Despite these numbers, only two meteorite falls were recovered in North America during all of 1998. A 2.6-kg meteorite fell in Monahans, Texas, on March 22, and a rock weighing more than 64 kg fell near Portales, New Mexico, on June 13. We do not observe most meteorites that fall, so there must be many rocks from space waiting for a collector with a sharp eye.

Tales of Meteorite Falls

Bath County is a remote, thinly settled region in the heart of eastern Kentucky, nestled in the Appalachian mountains. This is mining country, mostly coal. Bath County's population today is about 10,000; the second-largest village, Salt Lick, has a population of fewer than 800. As in much of rural Kentucky, the population today is not very different from what it was a century ago.

On November 15, 1902, at 6:45 P.M., Mr. and Mrs. Buford Staton of Bath Furnace—a little settlement about 5 miles south of Salt Lick—were disturbed from their dinner by a rumble and a flash. In a letter to the American mineral hunter Henry Ward, Mr. Staton described:

> It was dark. I saw the light and heard the report. It came through the air; whizzing like a steam saw going through a plank. . . . The stone struck in the middle of our hard road and bounded away for about five feet to one side. The hole which it made in the road was about one foot long, nine inches wide; and five inches deep.

The stone itself weighed 27 kg.

The "light" and "report" he mentioned were no mere local events. In fact, people saw the meteorite streaking through the skies above Louisiana, Mississippi, Alabama, Georgia, Tennessee, Kentucky, and Ohio, as well—a path nearly 1000 km long. The locals said that when this meteorite hit, it sounded like the high explosives used in mining. Other descriptions compared it to cannons, or guns fired in succession, or a protracted rumbling like a heavy carriage traveling over a log road.

Modern listeners recognize these noises as *sonic booms* like those from supersonic jets. Meteorites enter Earth's atmosphere at supersonic velocities—indeed, 10 times faster—and leave a boom in their wake.

But all this light and noise from just a 27-kg stone? No. The following May, Jack Pegrim was hunting squirrels in the woods a couple of miles south of Bath Furnace when he came across a scar on a white oak tree. Looking further, he found crushed saplings and broken roots nearby, then a hole at the foot of a larger tree. In the hole was a 200-kg rock with the same composition as the stone found by the Statons. Its surface was black, as if it had been melted. The front of the meteorite showed furrows radiating from a point—the trains of molten rock flowing back from the front of the stone, indicating the orientation of the rock as it flew through the atmosphere.

That probably was not even the whole meteorite. Other trees in the woods were also knocked down, but no other stones could be found. Given the roughness of the terrain, that was no great surprise.

Some falls have been quite spectacular. Thousands of iron meteorites rained on Sikhote-Alin, Siberia, on February 12, 1947; a few of them dug craters as big as football fields. Fortunately, few people live in that part of Siberia. An even bigger fall hit near Jilin, Manchuria, on March 8, 1976. Several hundred stones, the biggest one weighing more than a ton, hit an area north of the city. They missed wiping out Jilin itself by less than 160 km.

When meteoroids like Sikhote-Alin and Jilin break up in the atmosphere and shower the ground they produce a *meteorite-strewn field,* an elliptical region that can be many kilometers long. Meteorites are found throughout the field, although the largest specimens usually land at the downrange end.

It is inevitable that, at some time in history, falling meteorites must have killed people and animals. Anecdotes from China, spanning the period from ancient times to the twentieth century, have suggested cases where many deaths were attributed to material falling from the sky. However, the evidence is not strong enough to be certain that the fatalities were indeed due to meteorites, as other more common terrestrial phenomena could have been to blame. Modern records do document farm animals being startled by falling meteorites, and it is generally accepted that the Nakhla, Egypt, meteorite of 1911 hit and killed a dog.

Houses have been hit relatively frequently in modern times. A 2-kg stone lodged in the ceiling of the Cassarino family home in Wethersfield, Connecticut, on April 8, 1971. Eleven years later, Wethersfield was hit again. On the evening of November 8, 1982, as Robert and Wanda Donahue were watching an episode of *M*A*S*H* on TV, they heard what "sounded like a truck coming through the front door." A 13-kg rock crashed through their roof, bounced off the carpet, hit the dining room ceiling, knocked over a chair, and finally came to rest under the dining room table. There was no question it was a meteorite; hundreds of people across New York and Connecticut had seen the fireball.

On the evening of October 9, 1991, thousands of football fans watching high school games up and down the U.S. East Coast saw a fireball streak northward across the sky. From Virginia to New York, parents and coaches with video cameras turned them upward to capture an impressive streak of light. These videos made it possible to trace the orbit of the meteorite—both where it came from and where it went.

Where did it go? Forty-four kilograms of extraterrestrial rock smashed into the rear end of Michelle Knapp's Chevy Malibu, parked in front of her house in Peekskill, New York (Plate 14).

Where did it come from? A mathematical analysis of the tracks captured on videotape traces the meteorite back into space and, indeed, all the way back to the asteroid belt.

Fireballs in the Sky

Because fireballs are so bright, they can be seen during the day as well as at night. When a fireball is observed, it is usually kilometers above the surface of the earth and often is visible over many states. If the meteorite making the fireball reaches the ground, it usually lands several hundred kilometers from the observer. Reports of a meteorite hitting just on the other side of the road or behind a neighbor's house are common, even though the object really landed far away. This is an optical effect that can be rather disconcerting for observers. However, if scientists obtain several observations from people viewing the fireball from different directions, they can calculate its true trajectory and the approximate location of possible meteorites.

Fireball Hunting

If you see a fireball, you should stand still and determine the angle of its path with respect to the horizon. Estimate the altitude where the fireball was first seen and where it finally disappeared. An object at the horizon has an angle of 0° and an object directly overhead has an angle of 90°.

Before moving you should also determine the fireball's direction by taking compass readings where the fireball first appeared and where it disappeared. Since fireballs occur unexpectedly, you are not likely to be carrying a compass. However, you can measure the compass bearings later if you mark your position and the position of the fireball with landmarks on the horizon. For example, if the fireball disappeared directly over a flagpole, make a note of that and measure the compass bearing from your position to the flagpole as soon as you can.

You should immediately write this information down, because it is often hard to recall later. Then fill out a fireball report by contacting the American Meteor Society (on the Web at www.amsmeteors.org). In addition to the measurements just described, you should include information about the color of the fireball, its brightness, and the time it appeared or disappeared.

Limit these reports to very bright meteors or fireballs, not the small meteors you can see on any clear night. In most cases, these small interplanetary dust particles completely burn up in the atmosphere.

Hunting for Meteorites in Dry Deserts

Meteorites are like pages in a book about the solar system's history. Some chapters are missing, however, which is why scientists are always scrambling to find meteorites, hoping to discover new types of samples that will complete the story. David Kring (coauthor of this chapter) and his colleagues often search in deserts, because meteorites found in arid regions are better preserved. Whereas a meteorite may crumble to dirt within a couple of years in a humid environment, it may survive for thousands of years in a desert. Consequently, searchers are more likely to find a meteorite in one of these regions than in a rainy area like the U.S. Pacific Northwest. In addition, the desert landscape has very little vegetation, so it is easier to see rocks on the ground than it would be in a swamp or forest. The outback of Australia, the Sahara Desert, and the American Southwest are favorite spots for meteorite hunters.

Recently Kring and colleagues have been working in the Gold Basin region of the Mojave desert in Arizona. This area receives very little rain and supports only a few scattered bushes and Joshua trees. During summer months, temperatures often sizzle around 110 °F. However, the air is so dry that nighttime temperatures, particularly during the winter, often plummet well below freezing. Cold winds howl across the valleys at breakneck speeds, often threatening to topple their camp. But the dry air is a blessing, because the star-studded skies are spectacularly clear. The Milky Way forms a brilliant white path across the sky nearly every night. There were few better places for seeing comet Hale-Bopp.

In this corner of the Mojave desert, Jim Kriegh, a retired engineering professor, found two small meteorite fragments while prospecting for gold in November 1995. He realized the samples were meteorites because he was part of a gold-prospecting group that Kring had addressed several years ago. Gold prospectors are constantly searching out-of-the-way desert regions where meteorites are likely to be found. They also use metal detectors that are perfect tools for finding most types of meteorites. After knowing what to look for, Kriegh found new meteorites in the Gold Basin area and near Greaterville, Arizona.

The two meteorite fragments from the Gold Basin were particularly interesting, because they were clearly part of a much larger object. Jim Kriegh, his friends John Blennert and Ingrid Monrad, and David Kring searched for additional samples. On many days, they covered huge swaths of desert without finding a single meteorite. It was often draining work as they battled the sun and wind. There was no shade to speak of, and the winds blew so hard that often they could barely hear the tell-tale signal on the metal detectors. However, in a 2-year period they found more than 3000 meteorite fragments in a 130-square kilometer area (Plate 15). Most of the samples were partially to wholly buried in the soil, so they would have been nearly impossible to find without metal detectors.

Each time they discovered a new sample, they wondered what secrets it held. What chapters of solar-system history lay stored inside the rocky fragments? Once the samples were in the laboratory, Kring found that the meteorites contain chondrules, those bits of rock produced in a fiery rain amidst a primitive cloud of dust and gas nearly 4.6 billion years ago. After the chondrules had accreted to a planetesimal and been buried, they were baked at temperatures of several hundred degrees Celsius, causing some of the rocky material to recrystallize. Working with Tim Jull, another scientist at the University of Arizona, Kring determined that the meteorite pieces were part of a 2- to 3-m-diameter fragment of the planetesimal. About 15,000 years ago, this fragment hit the top of the atmosphere with the energy equivalent of 10 to 1000 tons of TNT. As the meteorite plummeted through the atmosphere, the pressure of the collision eventually overwhelmed its strength and caused it to explode, showering northern Arizona with thousands of pieces of rock. What a spectacular sight that must have been.

Hunting for Meteorites in Antarctica

In the early 1970s, a team of Japanese scientists led by Keizo Yanai realized that Antarctica had all of a desert's advantages for the meteorite hunter and more. The glaciers literally keep the meteorites in deep freeze, often encased in ice, where they are preserved even better than in dry deserts. Even more remarkably, the motions of the Antarctic ice cap concentrate meteorites fallen over thousands of square kilometers into regions of blue ice only a few kilometers across.

The south polar plateau is at an altitude of more than 3 km and cold enough that the snow never melts. The constant accumulation of snow compresses the ice below, squeezing out the air bubbles that make ordinary ice look white and producing a deep blue color. Repeated snowfalls thicken the ice cap, and the weight causes the ice sheet to flow slowly downhill for hundreds of kilometers, often reaching the sea.

Some of the ice never reaches the sea, though, because the flows are blocked by obstructions, like the Transantarctic Mountains. In the shadow of these mountains, snow rarely falls. Instead, constant dry winds from the south scour the ice, evaporating it. Meteorites are left behind—an accumulated trove of extraterrestrial rocks carried from where they fell along the ice flow path.

Since 1976, scientists from around the world have gone onto the ice to collect meteorites (Plate 16). About once every 10 years, the Japanese sail to a remote region along the coast and set up a semipermanent camp where they conduct many experiments in Antarctic science. From there, the meteorite hunters drive out to the ice fields in large tracked vehicles. They live and sleep in these ungainly machines, venturing out on foot during the warmer parts of the day to collect samples. Each expedition can take up to 18 months, including wintering over during the coldest and darkest part of the year. Along with the discomfort, there is also significant danger. During the 1991 expedition, one of the giant tracked vehicles was lost in a crevasse. Fortunately, both the passengers and the meteorites were rescued.

Every year, on the opposite side of the continent, a much smaller American expedition arrives at the permanent U.S. base at McMurdo. (Guy Consolmagno, coauthor of this chapter, was a member of the 1996 team.) From the base, military aircraft fly them to remote parts of the Antarctic interior. A team of six to eight people sets up camp in small but sturdy tents near the meteorite fields. Each day, weather permitting, they slowly drive snowmobiles in formation across the ice, looking for dark rocks sitting on the blue surface. After six weeks of high winds, isolation, and intensive searching, the teams return to McMurdo and home. The hundreds of meteorites they find are kept frozen until they reach the Johnson Space Center in Houston. There, scientists process and distribute the meteorites in much the same way that the Moon rocks have been cared for since the Apollo program.

Antarctic meteorite hunters have immensely expanded the number of samples in our collections. As of 1998, the Japanese had collected 9613 meteorites and the Americans had 8412 samples from Antarctica. By contrast, only about 4000 meteorites have been found anywhere else in the world.

What Should You Do with a Newly Found Meteorite?

Meteorites are particularly valuable scientific specimens because they come from planetary bodies (mostly asteroids) which we have not yet sampled through space missions. Indeed, meteorites will be our only samples of many types of asteroids for decades, if not longer. So it is important that new meteorites be brought to the attention of research laboratories.

It is also important to get meteorites into a laboratory as soon as possible after they have fallen or been found. Weather, chemicals in the soil, and even the perspiration on your hands can contaminate the samples. Meteorites cannot hurt you when you pick them up, but you can hurt them.

A good science museum or university geology laboratory will be able to determine whether your rock is a meteorite. They may also conduct a series of chemical and mineralogical analyses that can determine the type of meteorite that you have found. This information helps determine where in the solar system the meteorite originally came from and how long it took to get to Earth. Once the meteorite has been analyzed, a description is published so that the entire world knows that a new meteorite has been found.

If you find a meteorite on your property, it belongs to you (at least in the United States—other countries have different laws). You may choose to keep the meteorite, sell it, or donate it to a museum. In any case, a fraction of any new meteorite should be preserved for scientific study. This is important for two reasons. In science, when a conclusion based on a set of analyses has been announced, it is necessary for other scientists to be able to repeat the measurements to confirm the conclusion. Second, it is likely that our analytical techniques will continue to improve, so we want to set aside a portion of any new meteorite for future scientific analyses. Scientists often have been able to analyze meteorite samples that fell well over 100 years ago, because portions of them were safely stored in museums.

One of the most spectacular of these samples is a meteorite called Chassigny which fell in 1815. For years it was thought to be just an interesting sample, more of an oddity than of any particular importance. However, because portions were so well preserved in museums, scientists with modern instruments and more knowledge discovered that the meteorite came, not from the asteroid belt, but from Mars. As described in the next chapter, Chassigny and other Martian meteorites are providing rare glimpses of the geologic evolution of our neighboring planet.

Rocks from the Moon and Mars

Unexpected Visitors

In 1938, Orson Welles stunned America with a radio braodcast of *War of the Worlds,* in which he reported that Martians had invaded Earth. Hearing these reports, many thousands of people panicked, poured into the streets, and tried to flee the East Coast, site of the attack. To their relief (and, later, anger), listeners soon learned that the radio broadcast was fiction, based on a story written by H. G. Wells shortly before the turn of the century. Radio drama had been mistaken for a real invasion.

War of the Worlds is fiction, but there is also a true tale of invaders to tell. Both Orson Welles and H. G. Wells would probably have been fascinated to learn that Martians had already landed on Earth. These Martians were not flesh and bones, however, but rocky fragments of our red neighbor: meteorites—on Earth—from Mars.

Thus far, we have 14 confirmed Martian meteorites (and possibly as many as 15). One rock from Mars fell in 1815, during the early morning hours, in the small French village of Chassigny. Another landed in 1865 in Shergotty, India. It, too, fell in the early morning hours, filling the air with detonations. About 40 stones, collectively known as the Nakhla meteorites, fell in Egypt in 1911. A cloud of material filled the sky and explosions again boomed across the land. One of these stones reportedly killed a dog. A fourth Martian meteorite, called Zagami, fell in Nigeria in 1962. The 10 other confirmed Martian meteorites were found long after they fell—six in Antarctica, two in the Sahara, and two more in Indiana and Brazil (see Table 6.1). Another possible Martian meteorite, recovered during a recent Japanese expedition to Antarctica, has recently been reported. While the first of these meteorites was recovered in 1815, it was not until the early 1980s that scientists realized this group of rocks came from Mars.

Likewise, it was not until the 1980s that we realized we had samples of the Moon in our meteorite collections. An American field team in Antarctica discovered the first recognized meteorite from the Moon, Allan Hills (ALH) 81005, in 1981. Actually, a Japanese team had collected a lunar meteorite two years earlier in Antarctica, but it was not identified as a Moon rock until after the discovery of ALH 81005. Scientists have now confirmed the discoveries of 11 lunar meteorites found in Antarctica. Three have been found outside that icy continent—one in Australia and a pair in the Sahara (see Table 6.2).

Table 6.1 The Martian Meteorites

TYPE AND NAME	AGE OF ROCK, MILLION YEARS	EXPOSURE TO SPACE, MILLION YEARS	TIME ON EARTH, DATE OR YEARS
Shergottites			
ALH 77005	180	3	200,000
Dar al Gani 476	800	1.2	<135,000
Dar al Gani 489	Unknown	Unknown	Unknown
EET 79001	Unknown	0.5	Unknown
LEW 88516	Unknown	3	<50,000
QUE 94201	140	2.6	300,000
Shergotty	180	3	1865
Y 793605	Unknown	Unknown	Unknown
Zagami	180	3	1962
Nakhlites			
Governador Valaderas	1300	10–12	Unknown
Lafayette	1300	10–12	Unknown
Nakhla	1300	10–12	1911
Chassignites			
Chassigny	1300	10–12	1815
Unique			
ALH 84001	4500	14–16	13,000

Table 6.2 The Lunar Meteorites

TYPE AND NAME	EXPOSURE TO SPACE, MILLION YEARS	TIME ON EARTH, YEARS
Anorthositic breccia		
ALH 81005	0.0025	9,000
Dar al Gani 262	Unknown	Unknown
Dar al Gani 400	Unknown	Unknown
MAC 88104*	0.04	210,000–250,000
MAC 88105*	0.04	210,000–250,000
QUE 93069*	0.15–0.5	5,000–10,000
QUE 94269*	0.15–0.5	5,000–10,000
QUE 94281	Unknown	Unknown
Y 791197	<0.019	<60,000
Y 82192*	9 ± 2	80,000
Y 82193*	9 ± 2	80,000
Y 86032	9 ± 2	80,000
Basalt and basaltic breccia		
Asuka 881757	0.9	<70,000
Calcalong Creek	0.2	<70,000
EET 87521	<0.01	<60,000
Y 793169	1.1	<50,000
Y 793274	<0.02	<20,000

*Fragments of the same rock.

How We Know Rocks Came from the Moon and Mars

Asteroids and comets have bombarded the Moon for eons, making lunar rocks very different from those on Earth. Most lunar rocks contain fragments of many different rock types, welded together by the heat of impacts to form a coherent rock called *breccia.* Many of the lunar meteorites are impact breccias. ALH 81005, for example, contains bright white rock fragments from the lunar highlands as well as dark fragments of basalt from faraway lava flows. This jumble of rock fragments points strongly to the Moon. In addition, the chemical compositions of the meteorites and their mineral constituents, like planetary fingerprints, are very similar to those measured in samples collected on the Moon by Apollo astronauts.

It was more difficult to identify the source planet of the Martian meteorites. Indeed, until recently, they were called *SNC meteorites* rather than Martian meteorites. The term *SNC* is an acronym based on the names of the meteorite specimens Shergotty, Nakhla, and Chassigny, each of which represents a different type of igneous rock. Of the

14 Martian meteorites now confirmed, 9 are shergottites, 3 are nakhlites, 1 is called a chassignite, and 1 is the unique specimen ALH 84001 (see Table 6.1).

The SNC meteorites are chemically and isotopically related, strongly indicating that they came from the same parent body. We know that they came from Mars for two basic reasons. First is their relatively young crystallization ages. Volcanic processes produced most of these rocks between 140 million and 1.3 billion years ago. While this seems ancient in human terms, it is relatively young on a planetary time scale. Volcanism requires an internally hot planet. These meteorites must have come from a large planet that stayed hot long enough to produce such young volcanic rocks. Earth, Venus, and Mars are large and hot enough to have recent volcanic activity (Fig. 6.1). Neither Earth nor Venus, though, is the source of the SNC meteorites. Because we have analyzed so many rocks from our own planet, we know that these meteorites do not have Earth's chemical and isotopic fingerprints. We also know enough about Venus' geology from Soviet and U.S. spacecraft missions to know that the meteorites probably do

Figure 6.1 Volcano on Mars; photograph by the Viking spacecraft. (Image courtesy of NASA)

not bear its planetary signature, either. Consequently, the most likely source of the meteorites is Mars.

When a Martian source for the SNCs was first proposed, based on these arguments, few scientists were convinced. The clinching evidence came from an entirely unexpected source. One of the SNC meteorites, EET 79001, contains minute traces of trapped gas from the atmosphere of its parent planet. When scientists extracted that gas and analyzed it, they discovered it was identical to the mixture of gases measured in Mars's atmosphere by the Viking spacecraft. Since each planet has a unique atmospheric composition, EET 79001 must have come from Mars. The other SNC meteorites are then linked through EET 79001 to Mars because they all contain the same chemical and isotopic fingerprints. The wonderful thing about these samples is that they are free specimens from our distant planetary neighbor. Because they fell from the sky, we obtained them without having to spend hundreds of millions of dollars to build a spacecraft to collect and return the samples. To understand Mars, we still need to use spacecraft to collect additional rocks from known locations, but the free samples sent to us as meteorites remain a scientific bonanza.

Most people expect Martian rocks to be red and are surprised to learn that the meteorites are really gray or greenish gray. This lack of red color may seem odd since you can see a red point of light when you look at Mars in the night sky. Also, the surface of Mars is clearly red-brown in images from the Mariner, Viking, and Mars Pathfinder spacecraft. However, it is the fine-grained dust that coats the surface of the planet that provides this color. The dust is red because it contains a high percentage of iron, which has been oxidized to a rusty red color. A closer inspection of the Viking and Mars Pathfinder images shows that the underlying rocks often have dark weathered surfaces. From our studies of Martian meteorites, we know that fresh surfaces of rocks on Mars are generally gray to green in color, not red.

How They Got to Earth

Meteorites from the Moon and Mars have had complex histories. Not only did they survive a fiery plunge through Earth's atmosphere, they also experienced the cataclysmic impact of an asteroid or comet on their parent planet. Such impacts launched the rocky samples from the Moon and Mars and directed them upward, much like water splashing when you throw a rock into a pond. Most of this impact ejecta fell back to the surface, forming a blanket of rock and dust around the crater. However, a tiny fraction of that material reached a speed that overcame the gravitational force of the planet and flew into space. Over time, these rocks migrated on a variety of orbits and eventually collided with other planets or the Sun. Some of them even collided with Earth.

Scientists can calculate when the impact that launched a meteorite occurred by determining how long it was in space. They measure radioactive isotopes produced in the rocks by bombardment from solar flares and cosmic rays. The abundances of these isotopes relate directly to the length of time the rock spent in interplanetary space.

Our collection of lunar meteorites was produced by many different impact events. At least 7 and possibly as many as 10 impacts were required to launch all of the lunar meteorites from the Moon. The oldest impact event that produced a lunar meteorite occurred approximately 9 million years ago. In contrast, ALH 81005 left the Moon only 11,500 years ago. It made a very quick transit to the Earth, requiring only 2500 years, after which it lay in the Antarctic ice for an additional 9000 years (Fig. 6.2).

Figure 6.2 Lunar meteorite ALH 81005; 1-cm cube for scale. (Image courtesy of NASA)

What about the Martian meteorites? One or more impacts launched the nakhlites and Chassigny about 10 to 12 million years ago. Most shergottites left Mars 2.5 to 3 million years ago, but one was ejected less than a million years ago. The unique Martian meteorite ALH 84001 left Mars nearly 16 million years ago (Fig. 6.3). The impacts that ejected these meteorites represent a small fraction of the total number of impacts that occurred on Mars. Even with this small set of samples, we know that a moderate-sized asteroid or comet hit Mars at least every few million years. That is something to keep in mind when thinking about the possibility of large impacts on Earth.

The size of an impact event needed to launch rocks from the Moon is not very large. Most estimates suggest that the craters produced by these impacts are ⅓ to 1 km in diameter—smaller than Meteor Crater in Arizona. There are hundreds of thousands of craters this size on the Moon, so it is a safe bet that a great deal of lunar material has rained on the Earth in the past.

Because Mars is larger than the Moon and has more then twice the Moon's gravitational pull, it takes a more energetic impact to launch rocks from Mars. To escape Mars, rocks must achieve a velocity of at least 19,400 kph. This is more than 5 times the speed of a rifle bullet.

Figure 6.3 Martian meteorite ALH 84001; 1-cm cube for scale. (Image courtesy of NASA)

Impacts which produced craters 5 to 15 km in diameter were probably required to launch the Martian meteorites (Fig. 6.4). Such impacts are larger than the one which made Meteor Crater but much smaller than many other impacts on Earth.

What Lunar Meteorites Tell Us about the Moon

There are several types of lunar meteorites; anorthositic breccias from the lunar highlands are the most common. The highlands are the bright white areas of the Moon you can see from your backyard. The white mineral plagioclase dominates anorthositic rocks on both thc Moon and Earth. Whiteface Mountain, site of most of the skiing in the Lake Placid, New York, area, is composed of white anorthosite. The rock is so bright that the mountain is visible from far away.

Anorthositic breccias in the lunar meteorite collection (and in the Apollo Moon-rock collection) are easily recognizable because they have brilliant white chips of anorthosite in a gray matrix. Not only do these rocks tell us about impact cratering of the lunar surface, they also tell us about the very early history of the Moon. Anorthosites have relatively low densities and rise toward the top of a mass of molten

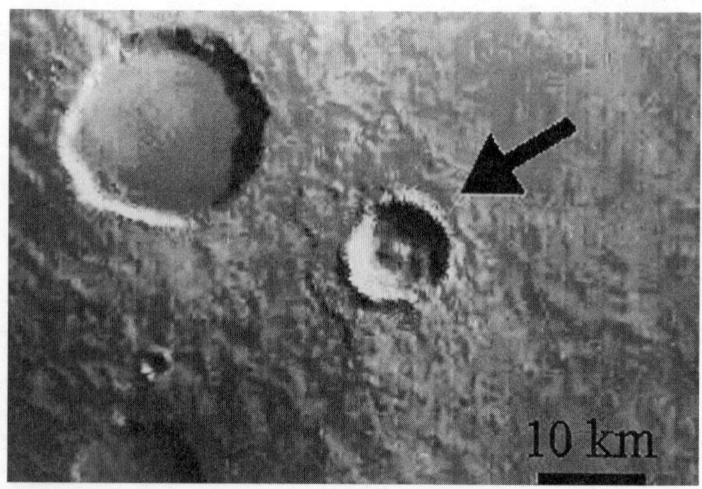

Figure 6.4 Possible source crater for Martian meteorite ALH 84001. (Image courtesy of NASA)

rock. Early in its history, when the Moon was either entirely or partially molten, anorthositic "rockbergs" floated to the surface, producing the bright white highlands that are visible today.

The lunar meteorite collection also contains a few samples of basalts and basaltic breccias. Basalt is the gray to black volcanic rock that makes up the dark areas you see on the lunar surface. Early astronomers thought these dark areas looked like seas, so they were given the Latin name for sea, *mare* (rhymes with "sorry"). The mare are not seas of water, but rather vast plains built of extensive lava flows. They are similar to basaltic lava flows in Hawaii. The basaltic lunar meteorites, together with samples in the Apollo collection, tell us that the Moon had extensive volcanic activity between 3 and 4 billion years ago. The combination of the lunar meteorites and Apollo samples is particularly important. While the Apollo collection is extensive and provides detailed information about specific locations on the Moon, the astronauts visited only six small regions. The lunar meteorites have helped us sample a much larger portion of the Moon's surface.

Scientists have long wondered about the Moon's origin. Because of its proximity, they have suspected it to be linked to the origin of our own planet. In addition, studies of lunar meteorites and Apollo samples indicate that there are many chemical similarities between the composition of the Moon and the composition of the Earth's mantle. But the exact relationship between our planet's origin and that of the Moon has long been uncertain. Some scientists believed the Moon accreted at the same time as the Earth from the same nebular material. Others proposed that it was once part of Earth but somehow was spun off. In the last two decades most lunar scientists have come to accept a catastrophic-impact origin for the Moon.

According to this theory, a Mars-sized planetary body collided with Earth early in our planet's history. This was, by far, the largest impact event ever to affect the Earth. The collision was so violent that it ripped off most of Earth's mantle, blasting a large fraction of this material, along with part of the colliding body, into a ring of debris that orbited the Earth. This shattered debris rapidly accreted to form the Moon. So when a lunar meteorite falls, a little bit of the Earth is being returned, spiced with material from a Mars-size planet that no longer exists.

What Martian Meteorites Tell Us about Mars

Because Martian meteorites are igneous rocks, they are a great source of information about the volcanic history of Mars. Each stone tells its own fascinating story, so this section describes some of them before summarizing what we have learned from the entire group. The descriptions of the Martian meteorites refer to pictures of them taken with a microscope (see the color insert section). Each picture shows a slice of the rock approximately 1 to 2 mm across. These are paper-thin (30-micrometer-thick) slices cut through the meteorites. Microscope thin sections are very useful types of samples for studying the history of a rock. Polarized light transmitted thorough different types of minerals and/or different orientations of the same mineral produces the brilliant colors seen in these photographs.

Plates 17 and 18, in color, show microscopic views of the nakhlite meteorites Nakhla and Lafayette, respectively, which are igneous rocks dominated by the mineral clinopyroxene. They also contain rare olivine crystals—one large orange-colored grain is in the upper right corner of the Nakhla picture. These rocks formed around 1.3 billion years ago when their crystals settled to the bottom of a magma chamber, a region of molten rock beneath the planet's surface. A chamber like this often feeds magma to volcanoes on the surface, where it erupts as lava flows. In addition to the original igneous minerals, traces of clay minerals were also found in Lafayette, indicating that water infiltrated the rock. These alteration products formed from interaction with small amounts of near-surface water, rather than a large circulating groundwater system. These meteorites were blasted off Mars by an impact 10 to 12 million years ago.

Plate 19 shows a microscopic view of the meteorite Chassigny, the only member of the chassignite group of igneous meteorites. This rock formed 1.3 billion years ago when grains of olivine crystallized underground and settled to the bottom of a magma chamber. The olivine sometimes surrounds minute crystals of water-bearing minerals, indicating water was in the parent magma and in the crust of Mars. Similar minerals exist in Earth rocks, but not in samples from the Moon. These mineral associations, as well as topographic features on the three planetary bodies, suggest that Mars once had much more water than the Moon, but less water than Earth.

Chassigny probably came from one of the young Martian volcanic provinces, which are located near the equator and in northern latitudes. The principal candidate sites are in the Tharsis region. Chassigny was ejected from Mars 10 to 12 million years ago. Because ages associated with Chassigny are similar to those of nakhlites, and the two groups of meteorites have similar chemical signatures, they were probably blasted off Mars by the same impact event.

Plate 20 shows a microscope view of a shergottite found in Antarctica, the meteorite Lewis Cliff 88516 (or LEW 88516). This meteorite is a relatively young basalt, produced by volcanic processes. LEW 88516 contains the minerals olivine, pyroxene, and plagioclase. The olivine and pyroxene are the gray and bright orange to pink crystals, respectively. Plagioclase was shocked so strongly by an impact that it changed to glass, which is completely black in this image. The shock occurred around 3 million years ago, when an asteroid or comet hit Mars and ejected the sample toward Earth. LEW 88516 landed in Antarctica less than 50,000 years ago.

Allan Hills 84001 (or ALH 84001) is a very ancient Mars meteorite composed mostly of orthopyroxene (Plate 21). The minerals crystallized and settled to the bottom of a magma chamber, forming a solid rock about 4.5 billion years ago. Sometime later, asteroids and/or comets bombarded the volcanic region. At least two impact events occurred nearby, fracturing the orthopyroxene and pulverizing some of it into ribbons or veins of broken crystals. Four billion years ago, fluids percolated through these fractures and deposited carbonate and sulfide minerals. These minerals contain microscopic forms, described in the next section, that some scientists argue are evidence of fossil Martian life. About 16 million years ago, another impact on Mars ejected this fragment of the planet's crust into space. Orbiting the Sun on a path that crossed Earth's orbit, the Martian rock collided with our planet 13,000 years ago, landing on an Antarctic ice field. Accumulating snow incorporated the rock into ice, which then flowed across the continent to the foot of the Transantarctic Mountains. When harsh winds reexposed the meteorite, scientists eventually discovered it.

One of the surprising lessons learned from the Martian meteorites is that the crust of Mars developed very quickly. The very ancient crystallization age of ALH 84001 indicates that the crust of Mars developed within 100 million years after planet formation. Other Martian meteorites prove there were at least two periods of volcanic activity, 1.3 billion years ago and 100 to 200 million years ago. This broad range of ages indicates that Mars had an extended history of volcanic activity, consistent with the numerous volcanoes seen in spacecraft images of the Red Planet. Minerals in the meteorites show that water-rich and water-poor fluids moved through the rocks in the crust of Mars. However, estimates of the amount of water involved are much lower based on the meteorites than are estimates based on spacecraft images of Mars's surface. Reconciling the two different estimates of water abundances on Mars is an important problem still needing to be solved.

While Chassigny, Shergotty, Zagami, and Nakhla all fell within the last few hundred years, almost all of the other Martian meteorites fell thousands of years ago. It is also clear that the orbital paths these groups of meteorites took to Earth are different. In the case of the shergottites, they made their way from Mars to Earth in less than 3 million years. The nakhlites and Chassigny took a much longer path that required 10 to 12 million years to make the trip.

Life on Mars?

The origin of life is one of the great mysteries of the universe. Although we do not know how life began, we know that it has thrived on Earth. That leads us to wonder whether life could begin somewhere else and, if so, whether it would be similar to life on Earth.

Mars is one of the few places in the solar system where life could have developed. The principal reason Mars has captured the attention of scientists looking for life is the evidence that it once had liquid water, which is a key requirement for all life on Earth.

Pathfinder and Viking pictures show no water on the present Martian surface (Plate 22.) There are no ponds, lakes, or flowing rivers in those images. Today, scientists believe that most of Mars's water is below the surface, much of it frozen in subsurface pore spaces. This layer of frozen water is called the *cryosphere* and is somewhat similar to the permafrost layer in the near-polar regions of Earth. Below the cryosphere, there may be a layer of liquid water. Although we have not detected it yet, instruments to find this deep water are being designed. While the bulk of Mars's water today is probably locked up underground, a small amount is found in the polar ice caps. In some of the Viking images, we can even see that there is enough water in the atmosphere to condense briefly and form frost on cold nights.

However, in the past, there was much more water on the surface of Mars. Orbital images show valleys formed by flowing water, huge channels produced by catastrophic flooding, and possibly the shorelines of ancient oceans.

The possibility of Martian life has fascinated humans for centuries, and recently we have begun the search in earnest. The Viking spacecraft that landed on Mars in 1976 carried instruments specifically designed to detect microscopic life in the planet's soil. They found no such indications at those first landing sites, but the idea persists that Mars may have once harbored some sort of living creatures.

A good place on Mars to look for life might be near the volcanoes. Volcanic activity is a good long-term source of heat, and this heat can melt ice to produce liquid water. It can also force the water to circulate through the crust, venting through hot springs and geysers. Studies of these features in Yellowstone National Park have taught us that such hot-water systems are excellent places for all kinds of microscopic life to develop and flourish. Ancient hot springs may be the types of environments where life on Earth originated. Images and spectra from orbiting spacecraft may soon reveal the sites of ancient hot springs on Mars.

Impact craters could be other good places to search for signs of life. Even small craters may be deep enough to penetrate a thin cryosphere, causing water to flood the crater and produce a lake. Dry lake beds on Earth preserve abundant evidence of the living things that once filled their waters. If an impact event is large enough, it can even melt rock in the target area. Some of this molten debris is ejected, but much of it pools inside the crater. It may take hundreds to thousands of years for the molten rock to cool. During that time, it can power hot springs just like a volcano.

The interaction of water with either volcanic or impact heat is less likely on Mars today than in the past. The Martian flood channels, widespread volcanism, and large impact craters all date from early in the planet's history. Consequently, a good place to look for signs of past life on Mars may be in the oldest terrains on the planet, the cratered highlands. While none of our spacecraft have landed in that type of terrain yet, Martian meteorite ALH 84001 is a sample from those ancient highlands.

In 1996, David McKay of the NASA Johnson Space Center and a team of scientists studying this oldest of Martian meteorites published a paper that stunned the world. They revealed that ALH 84001 contained what might be the first evidence of life beyond Earth.

ALH 84001 is an igneous rock, produced by the cooling of magma beneath Mars's surface. After the rock crystallized, a fluid percolated through it, depositing a new set of minerals. These are dominantly iron and magnesium carbonates—minerals associated with many marine and some terrestrial organisms here on Earth (Plate 23). When the team of scientists studied the Martian carbonates with a high-powered electron microscope, they saw provocative wormlike structures that they believed might be fossils of extremely small Martian bacteria (Fig. 6.5).

The team also performed a series of chemical analyses and detected traces of material that could be the remains of biologic activity. One group of chemical traces is called *polycyclic aromatic hydrocarbons* (PAHs), which are organic molecules that can be a by-product of bacterial decay. Unfortunately, many nonbiologic processes commonly produce PAHs on Earth. For example, you can find PAHs

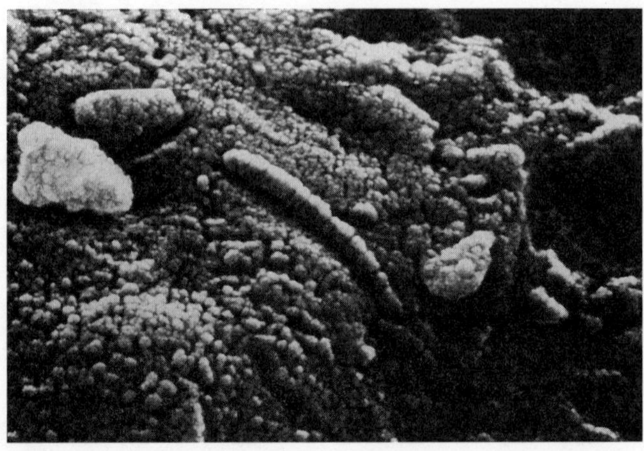

Figure 6.5 Possible fossil bacteria in Martian meteorite ALH 84001. (Image courtesy of NASA)

around a charcoal grill, a car exhaust pipe, or a burning oil lamp. So one fear was that the PAHs in ALH 84001 could be terrestrial contaminants. The team wrestled with this problem and found that the concentration of PAHs increased toward the interior of the meteorite. If the PAHs were contaminants, they argued, these chemicals should concentrate near the surface of the meteorite, not the interior.

The team also found extremely tiny crystals of iron oxide in the carbonates. Biologists had already discovered bacteria on Earth that produce iron oxide crystals identical in size, structure, and composition to the minerals in the Martian meteorite. This was another line of evidence pointing to possible ancient life in this rock from Mars.

Because of the potential importance of this discovery, many scientists immediately set to work attempting to confirm the claim of Martian life. Laboratories specializing in the mineralogy and chemistry of meteorites began examining ALH 84001 with increased interest. They were joined by many outside specialists who had never studied meteorites before, but who had expertise in a variety of fields associated with biology or the remnants of biologic activity. Dozens of papers have been published since the original proposal of fossil life was presented. Some investigators argue about the formation temperature of the orange carbonates—was it hot enough to destroy any organic material or cool enough for life to exist? Others have shown that the carbonates can form through normal geologic process, without the intervention of biologic activity. One group of scientists studying the PAHs contends that at least some of them could be terrestrial contaminants picked up in the Antarctic ice, while other investigators cite evidence that the PAHs are extraterrestrial.

The whole issue remains a compelling scientific detective story, as yet unresolved. In any case, the controversy has generated a series of fascinating investigations that have helped us learn much more about the geology of Mars and the chemistry of its rocks. We also understand much better how weathering on Earth can alter Martian meteorites, and we have developed a striking amount of new information about early life on Earth.

These studies are also an excellent prelude for a whole series of measurements that scientists from many countries plan to do on future spacecraft missions to Mars. Some of these missions will make measurements on the planet's surface, while others will bring samples back to Earth. In both cases, we need to develop better tools and methods for answering the question, was there—or is there—life on Mars?

Meteorites from Other Planets

Since we have found lunar meteorites and Martian meteorites, is it possible we may find meteorites from Mercury, Venus, or the satellites of the outer solar system? In some cases, the answer is yes. Impacts on these bodies can eject material into interplanetary space, and nothing prevents some of it from falling on Earth.

Mercury is a planet that we have not studied in much detail. No spacecraft has ever landed on its surface; we only have images from a flyby spacecraft and Earth-based telescopes. The surface is very likely covered with igneous rocks produced by volcanism and breccias produced by impacts. Consequently, the rocks jettisoned into Earth-crossing orbits probably include the same range of rock types that we have from the Moon and Mars. The compositions of these rocks will be different, but how different is uncertain. Some investigators have argued that they will contain very little iron and relatively high concentrations of aluminum, titanium, and calcium. A search of our current meteorite collections has turned up nothing unusual that can be linked with Mercury.

We know a little more about rocks on Venus because several Soviet spacecraft have landed on its surface. These missions suggest that igneous rocks with basaltic compositions are common, with some having a fairly high amount of sodium and potassium. A meteorite from Venus would probably contain pyroxene and feldspar, like the shergottites from Mars, although the composition of the feldspar would likely be different. Because Venus has a sulfur-rich atmosphere, a variety of sulfate and sulfide minerals could also be present. A Venusian meteorite might also be an impact breccia, because the surface of Venus is cratered. If we find a meteorite from Venus during the next couple of decades, it will likely be our first sample, since there are no current plans to build spacecraft that can bring back rocks from our sister planet.

We believe that rocky fragments from all of the terrestrial planets are traded back and forth, so it is possible for fragments of any one of the planets (and their moons) to be found anywhere. This has led some scientists to suggest that this trading of debris may be a way to transfer life from one planet to another. If we do find life on Mars, for example, careful study will be required to determine if it originated on Mars or arrived there on a meteorite from Earth. Likewise, we may want to ask ourselves whether it is possible that life first started on Mars and hitchhiked to Earth on a meteorite.

PART FOUR

Asteroids

Asteroids in Orbit

BY THE LATE 1700S IT HAD BECOME QUITE CLEAR TO MANY PEOPLE THAT stones can indeed fall from the sky. There were many eyewitness accounts of hundreds of rocks pelting entire towns or of hefty boulders landing with a thud in a farmer's field. Less obvious at the time, though, was the source of this aerial bombardment. Perhaps, it was thought, lightning strikes or tornadoes had hurled rocks from the ground into the sky. Maybe they had been blasted into the sky from volcanoes. In 1794, a German lawyer and physicist named Ernst Chladni published a paper in which he contended that meteorites were in fact pieces of extraterrestrial matter that had been swept up by the Earth. He even went on to speculate that they originated in celestial bodies that either had failed to form larger objects or that were created when larger objects had broken apart. Less than seven years after Chladni's farsighted ideas were published, the source of his celestial stones was discovered.

A Missing World

A Sicilian monk named Giuseppe Piazzi ushered in the nineteenth century with the discovery of the first asteroid, Ceres. Piazzi, who served as the director of the Palermo Observatory, was laboring to correct a star catalog when he spied a point of light in the constellation Taurus that was not on his star chart. Suspecting he may have discovered a new comet, Piazzi followed the object over the next few weeks as it moved slowly past the other stars. But Piazzi's "comet" did not display a comet's typical fuzzy appearance, and its motion was rather slow and uniform. Quite by accident he had discovered, in his words, "something better": the elusive quarry of a group of astronomers searching for a phantom planet.

Since the time of German astronomer Johannes Kepler's work in the early seventeenth century, many astronomers had suspected that a world awaited discovery somewhere in the gulf between Mars and Jupiter. Kepler, discoverer of the basic laws of planetary motion, was a devout believer in the harmony and grand design of the solar system. He felt that the 550-million-km emptiness beyond Mars had to be occupied by some planet.

In 1776 Kepler's countryman Johann Titius devised a simple numerical formula, later popularized by Johann Bode, which reproduced quite remarkably the spacing of the planets known at the time. According to the formula, a planet's distance from the Sun (relative to Earth's distance) could be found by first assigning a sequence of numbers to the planets: 0 for Mercury, 3 for Venus, 6 for Earth, 12 for Mars and so on, doubling each successive number for the next planet. Adding 4 to each figure and dividing the resulting numbers by 10 gives the distances of the planets from the Sun in astronomical units (1 AU is the distance of Earth from the Sun, about 150 million km): Mercury orbits at just under 0.4 AU, Venus at 0.7 AU, Earth at 1 AU, and so on. The formula works well until you pass beyond Mars. When you assign the next number in the sequence, 24, to Jupiter and run it through the formula, you get 2.8 AU, not Jupiter's actual distance of 5.2 AU. Aside from this one glitch of predicting a planet where none was seen, the formula worked well for the other planets known at the time, and it even successfully predicted the location of Uranus, discovered in 1781. (With the discoveries of Neptune and Pluto the Titius and Bode rule began to unravel, and most astronomers today consider it little more than a numerical coincidence.)

Titius and Bode's formula suggested that there ought to be a planet between Mars and Jupiter, and this led to the assembly of an international group of "celestial police" determined to find the missing planet.

Piazzi's chance discovery of Ceres in 1801 was the seeming culmination of the hunt, but the discovery of three other minor planets in rather quick succession shortly thereafter proved to be an embarrassment of riches. Those who had searched for Kepler's missing planet began to suspect a common source for the newfound worldlets. Were these asteroids remnants of some enormous celestial collision? Had the long-sought planet exploded, leaving only fragments to orbit the Sun?

Remnants of Planetary Formation

Today it is recognized that asteroids are not monuments to planetary death, but remnants of the birth of worlds. The formation of the planets was dominated by the collision and coalescence of rock and metal-rich scraps of material left over from the formation of the Sun.

As described in Chapter 1, our solar system was born in the collapse of a vast cloud of gas and dust nearly 4.6 billion years ago, perhaps triggered by shock waves from a nearby supernova. As the slowly rotating cloud contracted under its own gravity, it flattened into a disk, like a piece of dough tossed in the air by an experienced pizza maker. The central portions of the cloud continued to become denser and warmer, eventually shining forth as the Sun. Some of the gas and dust left over in the surrounding disk was incorporated into planetary bodies in a series of processes spanning about 100 million years.

At first, dust grains in the disk began sticking together, perhaps aided by electrostatic forces and their fluffy structures as they gently collided. Clumps of dust grains aggregated into larger particles, and these, in turn, accreted to form larger objects called *planetesimals,* some of them many kilometers in size. In the inner, warmer regions of the disk, planetesimals were composed only of those materials able to survive at high temperatures—rock-forming minerals and metals such as iron and nickel. Farther from the Sun, well beyond Mars in what today would be the outer regions of the asteroid belt, the temperatures were much cooler, and it is likely that significant quantities of water and other icy volatiles were incorporated in the planetesimals. Here, nature blurred the artificial line we have drawn between asteroid and comet.

A few planetesimals grew significantly larger than their neighbors. Their gravitational influence altered the motions of their smaller siblings, causing them to move in more eccentric orbits and to collide with ever-increasing speed. Collisions among the smaller planetesimals ejected fragments at speeds too great to be overcome by their weak gravity. Only the largest planetesimals could retain the debris of impacting objects. They continued to accrete, with the few successful survivors growing into the planets we see today.

This same process occurred in the region we now know as the asteroid belt. Here, however, where we might have expected another planet, the influence of mighty Jupiter altered the fate of the evolving planetesimals. As Jupiter began growing to its immense size, gravitational perturbations so altered the motions of the young asteroids that erosive collisions dominated over accretionary ones, and no planetary mass could ever gain a real foothold. Much material was so perturbed by Jupiter's influence that it was removed from the vicinity of the asteroid belt altogether, either by collision with Jupiter or ejection from the solar system. Today, what remains of the planet that might have been are a few large asteroids, including Ceres, Pallas, Juno, and Vesta, and a multitude of leftover planetesimals.

Myriad Mountains and Molehills

Most asteroids orbit the Sun in the main belt between Mars and Jupiter along paths that are fairly circular, matching the average orbital eccentricity of the major planets. Although the majority of the asteroids travel in orbits inclined less than 20 or 30° to Earth's orbit, a few make dizzying excursions above and below the plane of the solar system.

When the orbit of an asteroid has been well determined, that asteroid is assigned a number and a name. The number is a running index indicating its order in line of orbit determinations, and the asteroid's name as customarily suggested by its discoverer. The full designation for a minor planet therefore contains both the number and the name: 1 Ceres, 243 Ida, 4179 Toutatis, and so on. Discoverers have chosen names ranging from Greek and Roman mythology to the names of their spouses and friends, from cities to favorite pets. There are asteroids named for the Beatles, the seven Challenger astronauts, writers, scientists, artists, desserts, and flowers.

At 913 km across, Ceres is the largest asteroid, followed by Pallas and Vesta, with diameters of nearly 500 km. There are only a handful of asteroids larger than 200 km; as we look to smaller sizes, the number of asteroids increases rapidly. More than 9000 asteroids orbits have been well enough determined for them to be numbered and named. More than 220 asteroids are larger than 100 km, and from careful surveys of the sky, we estimate that there are nearly 10,000 larger than 10 km in size, more than 752,000 larger than 1 km, and nearly 28 million larger than a football field. It would seem that Kepler's empty expanse between Mars and Jupiter is quite crowded.

It is tempting to picture the main belt of asteroids as a sort of vast cosmic rock tumbler, with a multitude of orbiting mountains jostling against each other. We must remember, however, the enormous scale of the solar system. The planets are so small in comparison to the great distances between them that we rarely see properly scaled representations of the solar system. Few bookstores would be willing to stock a book thousands of feet across so that the planets could be drawn large enough to be seen, nor is it within the capabilities of the printing process to plot the orbits of the planets on a normal-sized page and then reproduce the planets as microscopic motes of ink.

A more practical way to grasp the actual scale of the solar system and appreciate the true emptiness of the main asteroid belt is by the use of a scale model. If we represent the Sun by a common 12-in-diameter globe, then the Earth would be about the size of a BB, orbiting just over 100 ft away, a mustard seed at 160 ft stands in for Mars, and Jupiter would be a ping-pong ball more than 550 ft from the globe. Since all of the asteroids combined would amount to a pile of rock less than half the size of the Moon, we can picture the asteroid belt in this way: Crush a large grain of sand into dust and scatter the remains randomly around you as you walk in a circle around the globe nearly 300 ft away. The seemingly countless millions of asteroids orbit in a vast volume of space. Movie heroes in their spaceships may have to dodge swarms of asteroids on the silver screen, but they would have no trouble at all navigating through *our* asteroid belt.

The distribution of orbits in the main belt is not uniform, and not all asteroids remain forever in the main belt. In 1867, a mathematics professor at the University of Indiana noted that there are gaps in the distribution of asteroid orbits at several locations in the main belt. Daniel Kirkwood interpreted these gaps, now named in his honor, as due to resonances with the motion of Jupiter, a kind of planetary no-man's land where gravitational perturbations from the massive planet sweep the belt free of asteroids.

Kirkwood explained that the gaps occur where an asteroid would have an orbital period which is some regular fraction of Jupiter's orbital period. Imagine, for example, an asteroid located at 3.28 AU from the Sun. Living closer to the Sun than Jupiter, the asteroid completes its orbit more quickly than the king of the planets, twice as fast in this particular case. For every one trip around the Sun made by Jupiter, the asteroid makes two. We refer to this as a *2:1 resonance.* On every second trip around its orbit, the asteroid finds itself near Jupiter, and feels a little extra tugging from the planet's gravity. Children very quickly learn that even a small kick, at just the right time, can pump up quite a ride on a swing. Kirkwood reasoned that in a similar way the repeated perturbations from Jupiter would, over time, alter the path of such an asteroid enough to remove it from the resonant orbit. Thus, a small region in the main belt is cleared of asteroids having average orbital distances which would place them in the resonance. Similar gaps occur at the locations of other resonances, such as 3:1, 5:2, and 7:3.

Some resonances actually collect asteroids rather than shoo them away. The most important of these is the 1:1 resonance, where asteroids orbit with the same period as Jupiter. In 1772, Joseph-Louis Lagrange wrote an "Essay on the Problem of Three Bodies," in which he suggested that the gravitational interplay between the Sun and a planet would create two regions 60° ahead of and behind the planet where a third, smaller body could find a stable orbit. Jupiter's *lagrangian points* are islands of gravitational stability in a sea of planetary perturbations and are home to the Trojan asteroids, a population of dark asteroids whose numbers rival those of their main-belt cousins. The Trojans are not tightly confined to the lagrangian points, but instead oscillate loosely around them, circling the Sun on their own independent orbits, remaining clear of Jupiter's embrace.

"Starlike"

While continuing to count the asteroids' numbers and follow their orbital escapades, astronomers have employed a wide variety of observational techniques to characterize the physical nature of the minor planets. As their name implies, asteroids appear as starlike points of light in all but the largest telescopes. But astronomers have become quite adept over the years at coaxing an impressive amount of information about celestial objects out of the light captured by telescopes. It is quite remarkable when you think of it—we can deduce the size, shape, composition, and a host of other details about a mountain of rock hundreds of millions of kilometers away without ever traveling to it or touching it, simply by collecting sunlight reflected from its surface.

The sizes of asteroids can be determined in a number of ways. If we can make out the disk of an asteroid in a telescopic image and we know its distance, we can immediately calculate its size. This technique is limited to the largest asteroids however, even with an instrument as capable as the Hubble Space Telescope (Fig. 7.1). A slightly less direct, but quite accurate, technique is based on the geometry of shadows. As we watch an asteroid wend its way across the sky, it will occasionally pass in front of, or *occult,* a distant star. By knowing how fast the asteroid travels in its orbit and by carefully measuring the blinking of the star's light as the asteroid briefly blocks it, we can determine the asteroid's dimensions. If many observers are located across the occultation track, we can gather information on its shape as well. In late May of 1983, Pallas occulted the bright star 1 Vulpeculea and the shadow passed across the southern United States, allowing a large number of observations. Although clouds obscured the southernmost portions of the occultation track, the data gathered to the north showed Pallas to be somewhat flattened in profile, with an average diameter of about 533 km (Fig. 7.2).

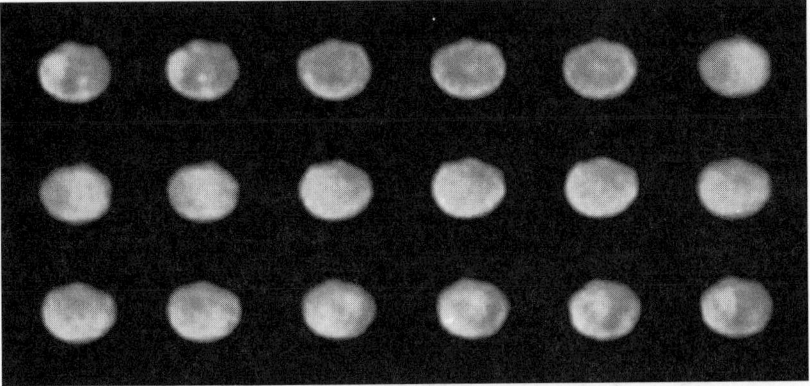

Figure 7.1 Mosaic of images showing the rotation of the asteroid Vesta; photograph by the Hubble Space Telescope. (Image courtesy of NASA)

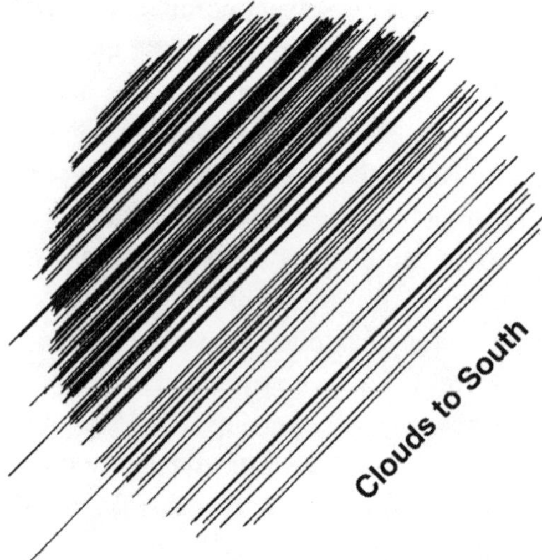

Figure 7.2 The outline of Pallas determined from occultation observations as the asteroid passed in front of the star 1 Vulpeculea; each line represents an observation from a different place on Earth. (Image courtesy of NASA)

We can also learn the shape of an asteroid by observing how its brightness fluctuates. If an asteroid is not perfectly round, but has an oblong shape, it will present to us faces of various sizes as it rotates. At one point we may be looking at the narrow end of the asteroid, while a quarter of a turn later, we will see its broad side. Of course, the apparent brightness of the asteroid depends on how much sunlight we see reflected from its surface, so when the narrow end of the asteroid is turned toward us, the asteroid appears dimmer than when the broad side faces us. This cyclic variation in apparent brightness changes as the asteroid and the Earth orbit the Sun and the geometry of our view of the asteroid changes. During one portion of its orbit, we may view the asteroid's equator, and its light-curve amplitude may be quite extreme, while several months later we may see the object's pole, when its light curve changes very little. When carefully analyzed together, many different observations yield the three-dimensional shape of the asteroid.

An asteroid can appear bright for two reasons: It may present a large area to reflect sunlight, or its surface may be composed of bright, reflective minerals. From half a solar system away, how can we tell whether we are looking at a big, dark asteroid or a small, bright one? Dark surfaces absorb more of the Sun's light than do bright ones. That's what makes them look dark—they do not reflect as much visible light back into space. But the excess energy they absorb warms the surface and is reradiated into space as infrared light. To our eyes, sensitive to visible light, the object looks dark, but in infrared observations, the object looks quite bright. By comparing its visible and infrared brightness, astronomers can determine the average *albedo* (reflectivity) of the surface of an asteroid. An asteroid that is bright in the visible part of the spectrum but dim in the infrared must reflect a lot of sunlight and have a bright surface. If it shines dimly at visible wavelengths but is bright in the infrared, its surface must be dark. With the help of the Infrared Astronomical Satellite (IRAS), more than 2000 minor planets have been measured. Many asteroids tend to be rather dark, reflecting less than a quarter of the sunlight that falls on them, and there is a strong tendency for the darkest asteroids to be found farther from the Sun. Most asteroids in the outer main belt are darker than charcoal.

Sunlight reflected off an asteroid's surface is changed in brightness and color. These changes provide information on the asteroid's composition. The reflectance spectra of asteroids show the presence of a number of common rock-forming minerals and generally match the spectra of meteorites that we have studied in detail (see Chap. 5). Astronomers have classified asteroids into a veritable alphabet soup of types based on the compositions implied by their various spectra. Most common in the inner main belt are the S types, which are composed of silicate minerals like olivine and pyroxene, with nickel-iron metal mixed in as well. They resemble olivine-dominated stony irons and some ordinary chondrite meteorites. The M-type asteroids have reflectance spectra dominated by the presence of nickel-iron metal and are thought to be the cores of minor planets. Most abundant of all are the dark C types, which dominate the outer asteroid belt. They most resemble the carbonaceous chondrite meteorites and are composed of silicate minerals with carbon in various forms. Many of the silicates found in these asteroids have water bound into their crystal structures and the spectra of some C types (Ceres in particular) even show the presence of water ice. Representing the rest of the alphabet are the black Ds and Ps, common in the outermost main belt and even more carbon-rich than the C types, and some less common types like the Vs, which look like basaltic lava; the nearly pure olivine A types; and the Es, Qs, and Rs, composed of the same silicates as the S types, with varying amounts of nickel-iron metal.

We have been able to gather quite an abundance of information about the general properties of the asteroids from these and other indirect techniques, but nature has kindly arranged to send us free samples of the asteroid belt for close-up scrutiny in the form of the meteorites. To do that the solar system has had to make molehills out of mountains.

Bashed to Bits

As we pointed out earlier, the asteroid belt is actually a rather empty place. There may be countless millions of orbiting mountains of rock out there, but the volume of space through which they travel is so vast that collisions between major asteroids are rare. But, given enough time, crashes can happen. And time is one thing that the solar system has plenty of.

In 1918, Kiyotsugu Hirayama noted that some asteroids had orbits that were quite similar to one another. Hirayama suspected that the individual asteroids in these groups were somehow related to each other and called the groupings, appropriately enough, families. He wrote that the existence of a family seems logical if you imagine a large asteroid being broken into many pieces, most likely as the result of a catastrophic collision with another large asteroid. The fragments would remain in the orbit of their larger parent body and continue to circle the Sun in orbits very similar to each other. Hirayama had discovered direct evidence of the most important process dominating the life of every asteroid—collisions.

The very gravitational stirrings responsible for the existence of the asteroids in the first place has condemned them to a life of perpetual pummeling. As their orbits cross, the asteroids approach each other at speeds around 5 kps. There are many more sand-grain-sized rocks and boulders out there than flying mountains, so for most of their lives large asteroids see little more than cratering impacts and smaller microimpacts which disturb their surface soil. But on rare occasions they cross paths with brethren nearer their own size. From impact experiments and mathematical analysis, we can predict the results.

When an asteroid is struck by another roughly one-tenth its size or larger, internal shock waves reflect off the opposite surfaces, producing strong tensile waves which fracture the target into many smaller pieces. The largest remnant may be only half the size of the original target asteroid. Debris ranging in size from kilometers across to motes of dust is ejected from the scene of the collision at speeds up to hundreds of meters per second. These ejection speeds are small, however, compared to the orbital speeds of asteroids around the Sun. While the newly formed fragments travel in their own independent orbits around the Sun, each follows an orbit closely resembling that of the original parent asteroid. This is how Hirayama's asteroid families are born.

At first, a new family exists as a cloud of rocky debris, but after only a few hundred years their relative speeds spread the young asteroids into a ring of sorts around the orbit of the original parent. Over a period of a million years or so, the orbits are further shuffled by perturbations from the planets, primarily Jupiter. Today, the individual members of a given family may be found on opposite sides of the asteroid belt, but they are recognizable as related siblings by their similar orbits. Hirayama found the three most prominent families by examining plots of the orbital elements of 790 asteroids known at the time. These were the Koronis, Eos, and Themis families, each named for the first-discovered asteroid in the family. Today, the orbital elements of many thousands of asteroids are examined using sophisticated statistical analyses. Some two dozen families are now recognized, conspicuous reminders of the massive collisions which have filled the asteroid belt with so much rocky wreckage.

Calculations indicate that catastrophic family-forming collisions occur in the main belt every few hundred million years or so. The debris from these rare, large collisions acts as projectiles to break up other asteroids, creating even more small debris. Over the 4.5-billion-year history of the solar system, the result has been a cascade of collisions responsible for the enormous number of very small asteroids that exist in the main belt. Because the smaller asteroids are so numerous, small-scale collisions are quite common. Every day, somewhere between Mars and Jupiter, several house-sized asteroids are blown to bits, disappearing in expanding clouds of boulders and dust. More substantial asteroids are commonly pelted by small projectiles, leaving their surfaces scarred with impact craters and launching meteorite-sized rocks into space.

Sampling the Debris

Eventually, most of the collisional debris produced in the main belt is removed from that region, some of it while still in the form of large asteroid fragments, the rest after it has been further ground into dust.

Gravitational resonances, as we have already seen, play a major role in perturbing the orbits of asteroids, clearing them from the Kirkwood gaps and demarcating the inner boundary of the main belt. Resonances also do most of the work in delivering asteroid debris to the Earth. Imagine a family-forming collision, or even a smaller, cratering impact on a more modest sized asteroid occurring near one of these resonances. Some of the collision fragments may be ejected in the right direction and at high enough speeds to reach the location of the resonance. Once there, the eccentricities of the fragments' orbits are pumped up to higher and higher values. It usually takes only about a million years or so for such orbits to become so elliptical that the fragments begin to wander into the inner solar system and make close approaches to Mars or even the Earth. These close encounters can substantially alter the fragments' orbits, removing them from the resonance entirely. They may then spend millions of years wandering the inner solar system. Some are destroyed by further collisions as they continue to loop out into the main belt on their elliptical orbits. Others dive into the Sun or are ejected from the solar system entirely by close encounters with Jupiter. The remainder continue to make close approaches to the terrestrial planets until their luck (and sometimes ours) eventually runs out.

The larger of these objects are the *Earth approaching* or *near-Earth asteroids* (NEAs). The NEAs that cross the orbit of Mars and approach, but do not actually cross, Earth's orbit are called *Amor asteroids.* Other NEAs do indeed cross Earth's orbit and, therefore, can potentially collide with our planet. These *Apollo asteroids* are a fairly representative sample of the various classes of main-belt asteroids. Being derived from collisional debris, they are usually quite a bit smaller than most of their numbered main-belt cousins, the largest being about 8 km across. The cratered surfaces of Earth and the other terrestrial planets bear witness to the ultimate fates of many of these objects and the grave consequences that such impacts can have for the inhabitants of our world, as discussed in more detail in Chap. 9.

The smaller fragments, blasted off of NEAs or delivered directly from the main belt by the resonances, may fall to Earth as meteorites. As actual samples of asteroids, meteorites are very valuable geologic specimens, telling us a great deal about the surfaces and interiors of small worlds we must usually study from afar. Those delivered directly from main-belt debris reach Earth by the relatively straightforward resonance mechanisms described earlier. Asteroid fragments generated farther from the resonances can also find their way to Earth, but they take a slower, solar-powered route.

As small rocks orbit the Sun and are warmed by the light they absorb, they reradiate the excess energy in the infrared portion of the spectrum. Because nearly all objects in the solar system rotate in some fashion or another, and because it takes time for rock to warm up and reradiate, the warmest spot is on the afternoon side of an asteroid. When the infrared photons depart this side of the body, they carry momentum away with them, giving the object a little kick in the process. Larger asteroids never notice the minute force from this reradiation; but for house- or couch-sized rocks, it can become quite important in perturbing their orbits. If an object spins in a *prograde sense,* like the Earth, the reradiated photons push on it generally in the same direction as it orbits, and the object spirals outward on a slowly expanding orbit. But if the meteoroid is rotating in a *retrograde sense,* the photons act to retard its orbital motion, and it drifts slowly inward toward the Sun in an ever-shrinking orbit.

This heat-driven orbit change is called the *Yarkovsky effect,* after the Russian engineer who discovered it around 1900. It has gained renewed attention from asteroid researchers in recent years for the important role it may play in transporting asteroid fragments created in otherwise stable regions of the main belt to resonances where they can begin their long journeys to the inner solar system and Earth.

Sunlight is also responsible for removing the smallest, dusty debris from the asteroid belt. As sand-grain- and dust-sized particles orbit the Sun, collisions with solar photons steal a little of their orbital momentum and cause them to spiral inward toward the Sun. This induced drag on small dust grains is called the *Poynting-Robertson effect,* after the British and American physicists who first described it. Micrometer- to millimeter-size grains can be transported from the middle of the main belt to the Earth by this process in as little as tens or hundreds of thousands of years.

As these particles approach the terrestrial planets, especially Earth, they enter regions of orbital resonance, and there is a significant chance that they can be captured for short times in the resonances. This creates "bottlenecks" in the flow of dust toward the Sun and results in rings of asteroidal dust associated with the orbits of the terrestrial planets. The Earth's resonant dust ring is substantial enough that it has been detected by orbiting infrared observatories like IRAS and the Cosmic Background Explorer (COBE).

Unlike larger and heavier meteorites or asteroids, dust particles from the resonant ring and others drifting through from the main belt can be gently decelerated by Earth's atmosphere when they encounter our planet. Many are indeed severely heated in the process, but some survive their atmospheric capture unmelted and slowly settle through the stratosphere, drifting for weeks or months before reaching the ground. Every year the Earth accumulates another 30 million metric tons of cosmic material in this way. Chapter 4 describes how NASA U-2 research aircraft collect some of these dust particles during flights through the upper atmosphere. Often invisible to the unaided eye, the dust particles can still be analyzed with the same techniques as larger meteorites and mined for the precious information they bring to us from their parent asteroids far beyond Mars.

Asteroid Spacecraft Encounters

THE POWERFUL TECHNIQUES OF GROUND-BASED ASTRONOMY AND REMOTE sensing that astronomers have brought to bear in the study of asteroids have yielded a wealth of knowledge about these small bodies. In the 200 years since the discovery of the very first asteroid, we have progressed from mere recognition of their existence to a point where we understand the basics of their origin, collisions, and dynamic history. Our general understanding extends to their sizes, shapes, and compositions. But to understand the asteroids as worlds in their own right really requires a close-up look. No matter how much we think we know from ground-based observations, revolutionary advances in our understanding of other worlds are made by actually traveling to them. Each time we have journeyed to another planet, our perceptions of that world have changed dramatically, sometimes after the discovery of something surprising and wholly unexpected. Now that all of the major planets and the nucleus of comet Halley have been visited, the asteroids beckon.

Starlike No Longer

In mid-November 1971, the Mariner 9 spacecraft reached Mars, the first spacecraft to orbit another planet. Unfortunately, Mars was shrouded in one of its global dust storms. With little else to do while waiting for the dust to settle, controllers back on Earth turned the spacecraft's gaze toward Mars's moons, Phobos and Deimos (Fig. 8.1). Astronomers had long suspected that the two tiny moons were in fact asteroids, captured by the planet early in its formation, perhaps by an ancient, more massive and distended atmosphere. From Earth we could tell that Phobos and Deimos were dark and had spectra resembling the C-type asteroids. The Mariner 9 images showed two irregular, cratered worlds and provided our first tentative preview of what asteroids might look like up close.

In 1976, the Viking orbiters gave us much clearer views of the Martian moons and allowed us to study them in far greater detail. Both showed evidence of a significant lunarlike regolith, the layer of dust

Figure 8.1 The Martian moon Phobos; photograph by the Viking 1 orbiter. (Image courtesy of NASA)

and broken rock fragments created on airless worlds by the continual bombardment of meteorites. With this fine soil layer, the craters on both moons look like small craters on our own Moon—shallow, bowl-shaped depressions surrounded by raised rims of impact ejecta. These were the smallest solar system objects we had seen up close until this point, and it was not obvious that a regolith or recognizable craters would form on objects with so little gravity. But, in spite of their low surface gravities, Phobos and Deimos have managed to accumulate craters and a substantial regolith.

Perhaps the most important lesson we learned from the Martian moons is that the surfaces of two small bodies can be quite different in detail and yet appear nearly identical from ground-based observations. Whereas Phobos and Deimos show essentially the same global properties, such as color, albedo, and the way they reflect light from their surfaces at different angles to the Sun, they are distinctly individual. Phobos boasts an extensive system of pitted grooves associated with its largest crater and Deimos exhibits many patchy variations in brightness.

Phobos and Deimos provided a tantalizing preview of what we might expect from our first views of an asteroid. The detailed regolith processes on both moons, however, might be quite different from those acting on similar-sized asteroids. The Martian orbital environment in which Phobos and Deimos exist is quite different from the open expanse of the main asteroid belt. The trajectories of ejecta knocked off of the Martian moons are dominated by the gravity field of the planet and not by those of the diminutive satellites themselves. Phobos and Deimos might recapture debris sharing their orbits around Mars, perhaps substantially altering their regolith properties and cratering records. And finally, Phobos and Deimos are compositionally quite similar to each other and represent only one kind of asteroid. There are many other types out there as well, and it would be nice to see something different. For these reasons asteroid scientists anxiously awaited their opportunity to get to see a *real* asteroid. The long wait finally ended in October 1991.

Galileo and Gaspra

On October 29, 1991, the Galileo spacecraft, bound for an orbital tour of the planet Jupiter and its four large moons, made the first-ever flyby of a minor planet, skirting just 1600 km from asteroid 951 Gaspra at a speed of 8 kps. The encounter was made possible by tweaking the trajectory of the spacecraft as it passed through the inner part of the asteroid belt on its roundabout route to Jupiter.

Before the Gaspra encounter we knew some of the general properties of the asteroid from ground-based study. Gaspra appeared to be a relatively small asteroid, about 15 km across and fairly irregular, judging from its light curve. Spinning once every 7.04 hours, Gaspra orbits the Sun at the inner edge of the asteroid belt at about 2.2 AU. Its reflectance spectrum revealed Gaspra to be a relatively olivine-rich S-type asteroid. The region of the main belt in which Gaspra resides is densely populated with many other small, S-type asteroids, generally thought to be fragments of one or more larger parent asteroids smashed by catastrophic collisions. Astronomers expected Gaspra to be a collision fragment, rather than a small survivor from the earliest days of the solar system.

As Galileo approached Gaspra, it snapped a series of distant optical navigation images in order to pin down the asteroid's position relative to the spacecraft so that scientific instruments could be precisely pointed during the actual encounter. Before the encounter, astronomers could specify Gaspra's location in space to no better than about 200 km, and that was not good enough to guarantee that the asteroid would be captured in the images taken as Galileo closed in. By accurately measuring the asteroid's position relative to the background stars, the navigation team was able to reduce the uncertainties in Gaspra's position by nearly 10 times. To allow for the small remaining uncertainty in Gaspra's position and for spacecraft pointing errors, the imaging team had planned to take a mosaic of images around the expected location of the asteroid. The hard work of the navigation team had paid off. After traveling for hundreds of millions of kilometers, they brought Galileo to within only 5 km of the spot they had targeted. The navigation and instrument pointing were so good that Gaspra appeared in more of the images captured by Galileo's camera than expected—right in the center of the mosaic.

The amazing images revealed Gaspra to be an irregularly shaped object, with dimensions of 18 × 11 × 9 km (Fig. 8.2). Its outline is rougher than any other small solar system object that we had seen, but its shape is not that unusual for a small body born as a collision

Figure 8.2 Asteroid Gaspra; photograph by the Galileo spacecraft. (Image courtesy of NASA)

fragment. Some of the large concavities and facets of the asteroid are likely the very surfaces of fracture that carved Gaspra from its larger parent, while others may be the scars of very large impacts suffered by Gaspra after it began life as an individual object. If some of these large concavities are indeed impact scars, such enormous collisions would surely have shattered the interior of Gaspra. In fact, a number of grooves have been discovered on the asteroid, similar to those seen on the Martian moon Phobos. They are most likely fractures or gaps between large blocks within Gaspra, hinting at its fractured interior. Most of the grooves are less than 2 km long and around 100 m wide. The pitted appearance of the grooves suggests the presence of some regolith on Gaspra, appearing as though loose surface material has drained downward into cracks. The grooves are not quite as prominent as those on Phobos, but it might be that Phobos has less regolith than Gaspra, or that the asteroid is structurally stronger than the little moon.

There are plenty of smaller impact craters on Gaspra, ranging in size up to 1.4 km across. Some are sharp and distinct, indicating their relative youth, while others have the softer and more muted appearance of old age. Every time a large projectile strikes a small body like Gaspra and creates a crater, its impact jolts the surface of the asteroid and scatters the regolith, like striking the bottom of a shallow pan full of sand. Over time, the jolting effects of large impacts and the cumulative pelting by many smaller ones gradually soften and degrade the appearance of craters.

Color images of Gaspra showed that some of the freshest-looking small craters and areas around some ridges were a little brighter and slightly bluer than the average reddish surface of the asteroid. (We are speaking of very subtle color variations—to the untrained eye Gaspra and other asteroids appear bland and pale gray.) It appears that very recent impacts and downslope movement of dislodged regolith excavate fresher, previously buried, regolith that is bright and blue. Some sort of space weathering process must darken and redden more pristine regolith after it has been exposed to the harsh effects of radiation and micrometeorite impacts. A similar process occurs on the Moon. However, it is not yet clear if the same mechanisms are responsible on Gaspra since the asteroids exist in a different impact environment than the Moon and their surface mineralogies are quite different.

Probably the most important lesson we learned from Galileo's brief encounter with Gaspra is that in general our predictions about the asteroid's size, shape, spin, brightness, and color from ground-based observations were correct. That gives us confidence that the many techniques we have developed over the years for learning about asteroids from afar actually work.

Discovery and Surprise at Ida

On the final leg of its 6-year journey to Jupiter, Galileo passed one last time through the asteroid belt, speeding by a second asteroid, 243 Ida, on August 28, 1993. This encounter gave us a chance to compare and contrast the properties of two S-type asteroids. Although they belong to the same asteroid class, Gaspra and Ida are different enough for such a comparison to be revealing.

Whereas Gaspra is a relatively olivine-rich S-type, Ida is a much more typical representative of that class, with a reflectance spectrum showing stronger indications of the mineral pyroxene. Nearly 56 km long, Ida is 3 times larger than Gaspra and much more elongated. In fact, Ida is one of the most irregularly shaped objects yet imaged by a spacecraft, almost 4 times longer than it is wide (Fig. 8.3). Some places on Ida are 10 times farther from its center than others. Such an irregular shape, coupled with Ida's almost dizzying spin period of 4.6 hours, makes for an extremely interesting gravitational environment on the surface. A nonrotating spherical object of Ida's mass and volume would have a surface gravity about 0.001 that of Earth's. But for the real Ida, the surface gravity ranges between 0.0003 and 0.001 *g*. If you could stand on the surface of Ida and drop a rock, it would take between 13 and 25 seconds to fall to the ground depending on where you were standing. The lowest gravities occur at the maximum and minimum radii—at the maximum radii you would be farthest from the center of Ida and the outward force from the asteroid's rotation would be at its greatest; at the minimum radii there simply would be very little material attracting you toward Ida's center.

Figure 8.3 Asteroid Ida; photograph by the Galileo spacecraft. (Image courtesy of NASA)

Ida's irregular shape and rapid rotation also make the dynamics of rocky fragments ejected from its surface by cratering impacts very strange. Instead of following simple arcs as they would on a round and slowly rotating world, debris launched from Ida must make complicated loop-the-loop trajectories over the surface. Those rocks not blown clear off Ida make short hops across the surface of the asteroid while being tugged and pulled by Ida's complex gravity field as the lumpy landscape rotates rapidly beneath them.

Such complex dynamics played a critical role in understanding the distribution of crater ejecta and regolith observed on the surface of Ida. The largest regolith particles are a collection of blocks scattered mostly around the rotational leading faces of the asteroid (Fig. 8.4).

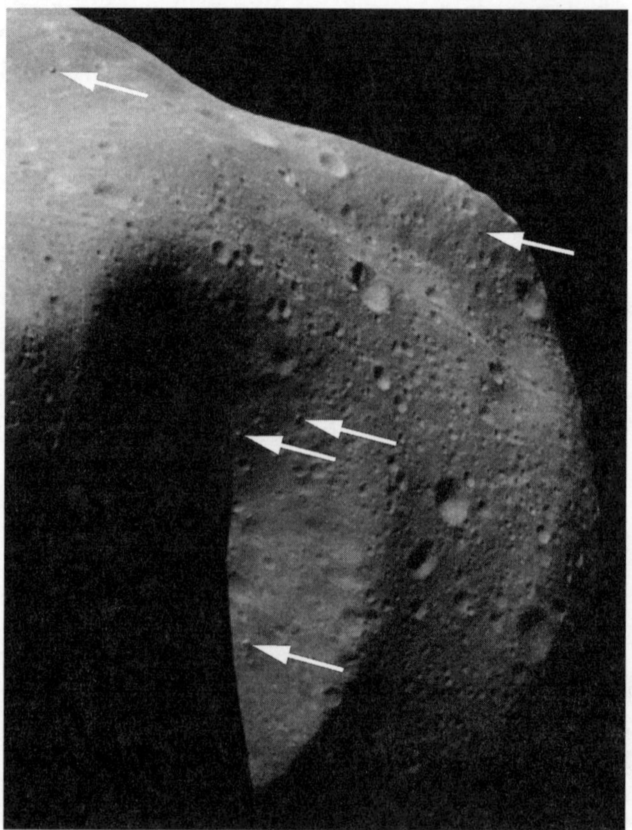

Figure 8.4 Closeup showing ejecta blocks (arrows) on Ida; photograph by the Galileo spacecraft. (Image courtesy of NASA)

Measuring 150 m across, these boulders are probably the largest fragments ejected at lower speeds during the excavation of an enormous crater. Computer simulations of the trajectories of such fragments show that they can be lofted high enough above the surface of Ida for the asteroid to complete a rotation or two before they fall back to the surface. Because of the rapid rotation of the asteroid, the elongated ends of Ida sweep through the hail of slowly falling fragments, preferentially collecting large ejecta blocks.

Blocks sitting exposed on the surface of Ida are rapidly destroyed by the continuous rain of small projectiles that pelts all asteroids in the main belt. Even the 150-m boulders will be smashed into rubble in tens of millions of years, so they must be from a relatively young crater. As it turns out there is a fresh 10-km-wide crater named Azzurra on the trailing face of the far side of Ida; the best candidate for the source of the blocks.

Imaged at lower resolution, Azzurra appears brighter and bluer than the rest of Ida, leading us to believe that the crater and the similarly colored material sprayed out around it have not been altered much by their exposure to the space environment. The same computer models used to predict the locations of large blocks ejected from Azzurra can be applied to the much finer debris ejected out of the crater at higher speeds. These models predict that fine ejecta blasted out of the crater lands right where we see the brighter and bluer regolith distributed across the surface of Ida, showing that the impact which formed Azzurra truly did shower Ida with a new contribution to its regolith, and confirming that some process of space weathering does indeed act to change the spectral properties of freshly exposed minerals from the asteroids' interiors.

The craters on Ida are named after various caves and grottos, since, in Greek mythology, the infant Zeus was hidden for safety in a cave on Mount Ida on the island of Crete. With its relatively bright, blue color, the crater Azzurra is well named after a flooded cave, accessible only by boat and known generally as the Blue Grotto, on the island of Capri. Other craters on Ida are named after the discovery site of Peking man, long lava tubes in Hawaii and Australia, a pristine and recently discovered cave in Arizona, and a beautiful blue-green spring in Florida.

Undeniably the most exciting and surprising discovery during Galileo's brief encounter with Ida was a small moon orbiting the asteroid. The diminutive satellite, named Dactyl, is a rounded object about 1.5 km across that is very similar in color and brightness to Ida. Astronomers had long wondered if asteroids could have satellites, and there were many observations hinting at their existence, but none could be confirmed. Dactyl's unexpected debut in Galileo's images of Ida decisively settled that question.

How did Dactyl form? Its larger companion, Ida, is a member of the Koronis asteroid family and is therefore a fragment born in the catastrophic collision that created that family. Computer models have shown that a few fragments of the parent asteroid could have ended up in gravitationally bound pairs, with one fragment orbiting another. Although the Ida-Dactyl pair may have formed in this way, Ida appears to be at least a billion years old. We do not expect an object as small as Dactyl to survive in the asteroid belt for much more than a few hundred million years before being blown to bits by an impact. Maybe Dactyl started out as a larger object and has since been whittled down by collisions. It is also possible that the asteroid formed in orbit around Ida, accreting from debris blasted off of Ida in a very large cratering event. We have already seen that large impacts can launch debris into temporary orbit around Ida, but we need to understand what happens if some of that debris collides with itself while in flight. Can some of it coalesce into a small body in a stable orbit? If Dactyl is indeed a very young object, formed long after Ida, could it have been born after the impact that excavated the crater Azzurra and scattered the blocks across Ida's surface? However Dactyl was formed, its discovery is one of the brightest jewels in the crown of achievements from Galileo's asteroid encounters.

From Distant Flyby to NEAR Rendezvous

Galileo spent only a few hours in the vicinity of Gaspra and Ida, hurrying past each asteroid to keep its appointment with Jupiter. For all that we learned about Gaspra and Ida, flyby encounters have their limitations, and many questions about each asteroid remain. For safety's sake, Galileo kept its distance from the asteroids, minimizing the risk of a mission-ending collision with the debris orbiting them. We could not directly measure the small masses of Gaspra and Ida through their gentle gravitational tugs on the spacecraft as it sped by. Galileo's instruments, optimized for study of Jupiter and its icy moons, were not designed to determine the elemental and mineralogical compositions of asteroids. And the ultimate limitation of any once-only flyby mission is that you cannot return to follow up in greater detail on unexpected discoveries.

The next natural step would be to fly a dedicated spacecraft to an asteroid, one designed to rendezvous with and orbit the asteroid for an extended period of time, equipped with instruments to study the asteroid in detail. In 1993, NASA announced that the first spacecraft of its new Discovery program would have just such a mission. The Near-Earth Asteroid Rendezvous (NEAR) spacecraft would be sent to orbit the Earth-approaching asteroid 433 Eros.

Like Gaspra and Ida, Eros is an S-type asteroid, but one that spends most of its time in the inner solar system, between Earth and Mars, rather than in the main asteroid belt. Eros falls within a class that provides the best match among asteroids to the reflectance spectra of ordinary chondrite meteorites, the most abundant type of meteorite that falls on Earth. Determining whether S-type Eros could indeed be a source of ordinary chondrites could solve a long-standing mystery of planetary science. Because its orbit lies so near that of Earth's, Eros is one of the easiest asteroids to reach, and being intermediate in size between Gaspra and Ida, it is large enough for a spacecraft to orbit. All of these characteristics make Eros a logical target for the NEAR mission.

The spacecraft was launched on a Delta II rocket from Cape Canaveral on February 17, 1996. NEAR's journey to Eros took just under 3 years, including a flyby of the main-belt asteroid Mathilde on June 27, 1997 and an Earth swing-by on January 23, 1998. The Mathilde encounter offered a golden opportunity to examine first-hand a class of asteroids that scientists had not yet seen close up. NEAR was going to pass very close to Mathilde anyway, and the spacecraft's solar panels would provide just enough power to operate NEAR's camera, so mission planners decided to have a look.

Unlike the S-type asteroids that we had already seen, Mathilde is a C-type asteroid, as black as charcoal and thought to have a composition like that of the carbonaceous chondrite meteorites. NEAR's camera revealed a tortured and battered world, 60 km wide and scarred with incredibly huge impact craters, some almost as large as the asteroid itself (Fig. 8.5). It is remarkable that Mathilde could have survived

Figure 8.5 Asteroids Mathilde, Gaspra, and Ida; mosaic of photographs by the NEAR and Galileo spacecraft. (Image courtesy of NASA)

one such colossal impact, let alone several, without being completely destroyed. The asteroid's interior at least should have been thoroughly broken up by such a pounding. NEAR flew close enough to Mathilde for the asteroid's gravity to affect the spacecraft's trajectory. By carefully tracking the spacecraft's motion, the science team could measure Mathilde's mass. With this mass, and knowing Mathilde's shape and volume, we calculated the asteroid's density to be about 1.3 g/cc, a little more than that of water and about half that of typical carbonaceous chondrite meteorites. If the asteroid's interior is porous, full of gaps and voids between rocky fragments, that could explain Mathilde's low density. Basically Mathilde is nothing but a pile of rubble, its lack of coherence allowing it to withstand repeated impacts.

In January, 1999 the NEAR spacecraft passed close to Eros and should have fired its rockets to slow down and rendezvous. A computer glitch prevented the firing and the spacecraft sped on past. Mission scientists will try for another rendezvous in 2000. Then for 1 year the spacecraft will orbit the asteroid, its suite of 5 specialized instruments scrutinizing Eros in great detail. As NEAR settles into its 100-km-high orbit around Eros, the multispectral imager will capture thousands of pictures through color filters. The images will be used to characterize Eros's shape, count craters on its surface, and study the nature and distribution of regolith. During the lowest of orbital passes, NEAR should be able to image individual large rocks scattered across Eros's lumpy surface.

The near-infrared spectrograph will measure sunlight reflected from the rocks and soil on Eros's surface and will map the distribution and abundance of the minerals that are found there. NEAR's spectrometers will detect the x rays and gamma rays that come from the asteroid's surface. This high-energy radiation is emitted after the soil is bombarded by cosmic rays and charged particles from the Sun and is characteristic of the elements present in the surface minerals. When coupled with color images, the information gathered by these instruments will provide a comprehensive survey of the elemental and mineralogical composition of Eros. Comparison with the compositions of meteorites may finally resolve the asteroid–meteorite connection.

Pulses of laser light bounced off of Eros's surface will make highly accurate measurements of the asteroid's shape and surface structure. The laser rangefinder is a very precise altimeter that will measure the distance between the asteroid and the spacecraft up to 8 times every second, complementing the shape information that we will get from the images of Eros taken from various angles. As the magnetometer maps Eros's magnetic field and the radio science team carefully tracks the spacecraft's motion and maps the asteroid's gravitational field, we will in effect peer below Eros's surface to deduce its internal structure, discovering whether the asteroid is a loose jumble of rubble like Mathilde or a solid slab of rock.

Mining the Sky

NEAR's rendezvous with Eros will provide our first close-up look at one of thousands of near-Earth asteroids (NEAs), objects of great interest to us for several reasons. As we have already seen, the NEAs are samples of main-belt asteroids and the immediate parent bodies of some of the meteorites that fall on our planet, delivering to us small but valuable samples of extraterrestrial material. NEAs are also the primary source of large objects that occasionally collide with our planet, fundamentally influencing the climate and the evolution of life, a topic covered in greater detail in the next chapter. Finally, some NEAs are, almost literally, gold mines in the sky, intriguing destinations for human exploration and a vast and as yet untapped source of mineral resources.

Many of the small asteroids that approach our planet are the most easily accessible bodies in the solar system, requiring a smaller total rocket thrust to reach them than would be needed for a trip to Mars, other planets, or even the Moon. Aside from the numerous compelling scientific reasons for mounting an expedition to an NEA, such missions will also serve as an excellent rehearsal for future Mars missions with human crews. We could use the NEAs as stepping stones to Mars, testing the capabilities of the hardware, probing the limits of mission design, and inspiring countless individuals with an exciting preview of a human flight to the Red Planet.

What would it be like to explore a near-Earth asteroid in person? Transit times from low Earth orbit could range anywhere from a few months to well over a year, depending on the particular target of exploration. Upon arrival at the asteroid, the spacecraft would initially enter a high orbit, and the crew would make a preliminary survey of the asteroid's shape and gravitational field. As the vehicle's orbit was lowered closer to the asteroid and knowledge of the gravitational field was refined, the crew would study the detailed shape of the rocky surface, map the distribution of minerals, and scout out safe but scientifically interesting landing sites. Once the spacecraft was parked in a stable orbit, it would serve as a base of operations for several excursions to the asteroid's surface over a period as long as two weeks.

The surface of a small asteroid will present astronauts with an environment quite different from the Moon or Mars. There will be myriad craters of various sizes and a dusty, rock-strewn regolith, but at this point any similarity to the lunar surface experienced by the Apollo astronauts ends. The topography in some places may be quite extreme. Saddles between overlapping large craters, ridges and facets defining the overall shape of the asteroid, and large ejecta blocks could make for a wildly varying terrain. On worlds the size of small cities, the horizon is never far away; the vista over the other side of the next hill might well fall away to encompass a whole other side of the asteroid. Astronauts will also have to deal with rapidly changing lighting conditions. The rotation periods of many small NEAs are short, and surface crews may well see several sunrises and sunsets during their excursions as they explore their tumbling mountain in space.

The very low surface gravities and rapid rotation of NEAs will make moving around on them quite interesting. In some ways the near-weightless environment will assist the astronauts—they will be able to reach just about any point on an asteroid from their orbiting spaceship, and their mobility on the surface will be enhanced as they drift over rough terrain and high obstacles. The surface environment will also pose daunting challenges, however. The astronauts will need very capable transportation and navigation systems for moving on the asteroid's surface. Personal transportation and maneuvering units, similar to those used by Space Shuttle astronauts, might be needed by the surface crew to transit back and forth between the orbiting vehicle and the asteroid's surface. When using the maneuvering unit's gas thrusters near the asteroid's surface, astronauts will have to be careful not to stir up dust that might disrupt equipment and scientific samples.

Once on the surface, the exploration team will survey and photograph the local environment and collect geologic samples. Instead of bounding about the surface as the Apollo astronauts did on the Moon, the crew in the microgravity environment of an NEA may float over its surface, tethered to the asteroid like astronauts repairing a satellite in Earth orbit. Asteroid surface teams will have more in common with cave divers than Moon walkers. The astronauts will enjoy all the benefits of three-dimensional motion, but they will have to move with care and finesse, perhaps at times using only a fingertip to move or steady themselves with their attached tools, instruments, and samples. After field work has been completed at one scientific site, the team will again don their maneuvering units, pull up stakes, and move to another site on the asteroid.

The first human missions to NEAs will concentrate on scientific studies, making invaluable measurements and returning precious scientific specimens to Earth for careful analysis. But ultimately it is the enormous resource potential of the NEAs that will make them valuable destinations for continuing missions. The NEAs are a compositionally diverse population of rocky and metallic bodies, offering us a mineralogical smorgasbord far richer and easier to reach than the materials present on the lunar surface. Almost all NEAs have at least 100 times more free metal not bound up in silicate minerals than Moon rocks do, and some of the more volatile-rich ones may contain 100 times more water than any lunar soils.

Nearly pure flakes of nickel-iron alloys could be magnetically raked from the regolith covering a typical chondritic asteroid. The by-products from the refining of these alloys, precious and strategic metals like platinum, palladium, chromium, and gold, will also be important resources. Abundant sulfur would be valuable, as would oxides of elements like calcium and aluminum, which could be used in making heat shields. Even the unprocessed regolith itself would be useful in providing bulk material for protective shielding from cosmic rays for longer human missions. Carbonaceous asteroids would be a rich source of water and organic compounds for life support and storable rocket propellants.

As the population of our small planet continues to grow and pressures to exploit ever-dwindling resources here on Earth increase, it seems only natural to begin to look elsewhere for some of the materials we will need. As our civilization expands into space and we become a multiworld species, a well-balanced program of asteroid research and use of asteroid resources may help our planet recover from the harsh treatment we have inflicted upon it.

Doomsday Asteroids

Impacts Throughout the Solar System

Before the Apollo program we had little idea of how important impact cratering is in the solar system. Even the Moon's craters were poorly understood, some scientists arguing that they were formed by impacts and others believing they were the result of volcanic processes. The issue was settled, however, when astronauts landed on the lunar surface and began collecting breccias shocked by levels of pressure that are produced only by impacts, not volcanoes. They also found rocks coated with glass, the tell-tale sign of impact melting. It soon became undeniable that impacts were responsible for most of the Moon's craters.

Lunar impact craters come in all sizes. Most people are aware of the 1- to 10-km-wide craters that we see in Apollo photographs (Fig. 9.1), or the large impact basins we can see from Earth. At the other extreme, some of the most amazing craters are smaller than the head of a pin. Micrometeorites slam into the lunar surface at velocities in excess of 40,000 kph. Because the Moon does not have an atmosphere, micrometeorites do not burn up as they do when they collide with Earth. They make microcraters that are 100 million times smaller than the great lunar basins, but it was the same process that formed them all.

Figure 9.1 Craters on the lunar surface; photograph by the Apollo astronauts. (Image courtesy of NASA)

The findings of the Apollo program prompted scientists to search seriously for impact craters on Earth. Researchers had previously described a few terrestrial impact craters, but only a handful of scientists were aware of their work. Not long after Apollo, the list of known terrestrial craters grew very quickly. Geologists conducted detailed field studies of these structures, while experimenters began simulating impact events in their laboratories. Our understanding of impact-cratering processes grew enormously.

As the result of more recent spacecraft missions, we now have evidence of impact cratering throughout the solar system. Cratering is clearly one of the dominant (if not the dominant) geologic processes on all of the rocky planets, the asteroids, and many planetary satellites. Mercury's surface, for example, is peppered with impact craters (Fig.

9.2). One of them has the enormous diameter of 1300 km, which is about one-fourth the diameter of the entire planet. Earth, Venus, and Mars also have impact craters, although less evidence of them is retained because many types of geologic processes obliterate craters. On Earth, erosion, mountain building, and plate tectonics destroy craters, while sediments and lava flows bury them. Venus, which is the same size as Earth and also geologically active, lost most of its impact craters to some of these same processes. Radar images show lava flows burying Venusian impact craters and faults tearing them apart (see Fig. 9.3).

Figure 9.2 Mercury; photograph by the Mariner 10 spacecraft. (Image courtesy of NASA)

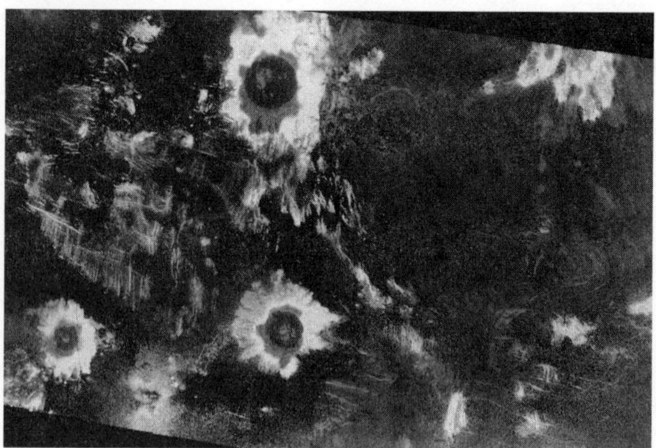

Figure 9.3 Craters on Venus; radar image by the Magellan spacecraft. (Image courtesy of NASA)

Ancient Martian craters are severely degraded, possibly by water and wind erosion.

While the giant gaseous planets (Jupiter, Saturn, Uranus, and Neptune) do not have the rocky surfaces needed to preserve impact craters, they have nonetheless been the targets of countless asteroids and comets. Comet Shoemaker-Levy 9 and its 1994 run-in with Jupiter is a spectacular example of such an impact. Geologists who study impact cratering were thrilled to witness such a rare large impact, particularly one we could view from a safe distance.

As the Shoemaker-Levy 9 event demonstrated, impacts affect planets throughout our solar system. While the giant planets' thick atmospheres do not preserve craters, most of their moons are visibly battered. Indeed, observations of those satellites indicate that breakups of comets like Shoemaker-Levy 9 have happened before. Impact crater chains on Jupiter's moons Callisto and Ganymede bear an uncanny resemblance to the imprint of the Shoemaker-Levy 9 string of comets.

Having surveyed the entire solar system (with the exception of faraway Pluto) with our spacecraft, we now understand the importance of cratering in every planet's and satellite's history. While the Earth may appear relatively unscarred, we know that our planet has also been the target of hundreds of thousands of impacts over the eons.

Earth Impact Craters

Studies of impact craters did not start until the late 1890s and early 1900s, when geologists and miners began examining Coon Mountain (now called Meteor Crater) in northern Arizona. The leading geologist in the country at the time, G. K. Gilbert, wrongly concluded that the crater was produced by some sort of steam explosion. In contrast, the mining engineer D. M. Barringer deduced that the structure was an impact crater, in part because the landscape around the crater was strewn with fragments of iron metal. Believing the bulk of an asteroid was buried beneath the crater, Barringer acquired the land so that he could mine this metal. Despite years of searching, he never found the mass of iron that he thought should be there, leading many to question whether the crater was made by an impact after all. At that time nobody understood that the energy involved in an impact is so tremendous that the colliding object is almost entirely obliterated. Metal fragments around the crater were all that remained of the original asteroid—there was no buried mass of metal to be found. Barringer was vindicated in the 1950s and 1960s, however, when Gene Shoemaker and his colleagues found high-pressure minerals and geologic structures which showed conclusively that an impact, rather than volcanism, had produced Meteor Crater.

The list of terrestrial craters has grown rapidly in the three decades since Apollo. Today there are about 150 impact structures identified on Earth, and each year we find around 5 more as exploration efforts continue. Some of these craters are relatively young while others are incredibly ancient. Haviland, Kansas, and Sobolev, Russia, may be the youngest craters; each formed less than 1000 years ago. The oldest crater identified thus far, Vredefort, South Africa, is 2.023 billion years old. Craters range in size from 15 m (Haviland, Kansas) to 250 km (Sudbury, Ontario). The smaller ones, like Meteor Crater, have relatively simple bowl shapes. However, most impact structures larger than 2 to 4 km have complex geometries that include an uplifted central peak or peak ring.

On both Earth and the Moon there are scars left not only by impacts from single asteroids and comets, but also double craters, suggesting the simultaneous impact of two projectiles. The lunar crater Bessarion B is a particularly good example, consisting of two craters (the larger 12 km across) with a ridge of material between them that was ejected during the violent instant of double impact. Back on Earth, at least three of the largest, well-preserved craters are doubles. Ries crater (24 km across) in Germany and its smaller 3.8-km-diameter sibling, Steinheim, formed nearly 15 million years ago. The 25-km Kamensk and 3.5-km Gusev craters in Russia date from about 49 million years ago. Probably the most striking examples are the East and West Clearwater Lakes in northern Quebec, where two asteroids of nearly the same size slammed into the Canadian granite some 290 million years ago (Fig. 9.4).

The fact that about 10 percent of the large craters on Earth are doubles tells us that a sizable percentage of the near-Earth asteroids must be binary objects. Bill Bottke of Cornell University and his collaborators have shown that very close encounters with Earth can disrupt weak asteroids in the same way that Jupiter's gravity tore apart comet Shoemaker-Levy 9. The remnants of these ill-fated asteroids often reaccumulate into binary objects or asteroid-satellite pairs rather than single objects. Upon subsequent encounters with Earth, these pairs may together impact our planet to form the double craters.

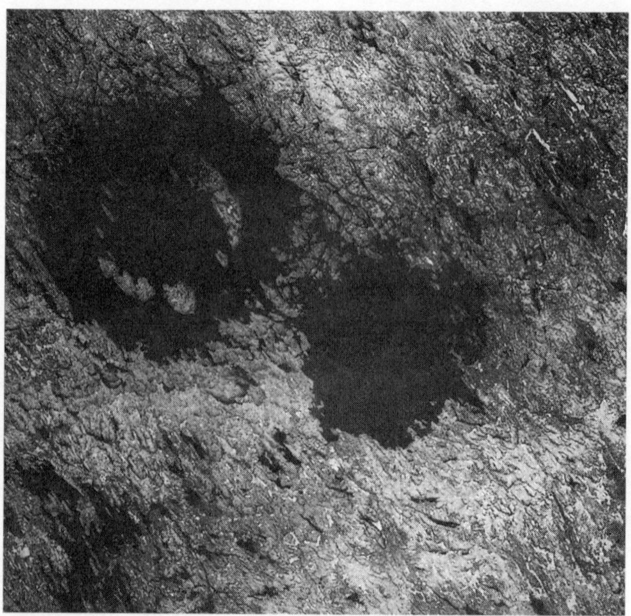

Figure 9.4 East and West Clearwater Lakes, Canada; photograph by the Shuttle astronauts. (Image courtesy of NASA)

The 150 known impact structures undoubtedly represent a small fraction of Earth's total count of craters. A much better sense of the battering Earth has endured comes from looking at the Moon. If our planet did not have internally driven geologic processes like volcanism and plate tectonics, or a climate that powers erosion, then Earth's surface would look similar to the almost unchanging face of the Moon, which has accumulated more than 300,000 craters of significant size. Imagine if we took away the grass, the mountains, the oceans, the eons of tectonic activity—Earth's surface unveiled would be peppered with craters, the mementos of its long days sailing through perilous space.

Tunguska, Russia

While the seemingly untouched face of our planet can make us forget, we were recently reminded of the hazards of impacting asteroids in 1908 when a blast rocked the cold forests and swamps of Siberia. When the day began, people woke to a clear sky and perfectly still air. At breakfast time, a man named Semenov was sitting on the porch of a house at a small trading post. His wife was inside, and his daughter was behind the house drawing water from a spring. Suddenly, Semenov saw something high above the forest that seemed to split the northern sky in two. He said the object covered the sky with fire while at the same time he felt a tremendous pulse of heat, as if his shirt had caught on fire. A loud bang was followed by a thunderous crash. Semenov was thrown to the ground about 6 m from the porch and lost consciousness. His wife and daughter pulled him into the house, and when he awoke, it sounded as if guns were firing. The ground was shaking and a hot wind blew past the house, scouring the ground. When things finally quieted down, he discovered several blown-out windows and a broken iron hasp on his barn door.

Elsewhere, the Potapovich family was sleeping in a tent. Suddenly, the tent blew up into the air with the family still inside. It then came crashing back to the ground, bruising everyone. Two of the people in the tent fainted. The ground shook as the air filled with a roaring sound. When everyone regained consciousness and looked outside, they saw smoke and found that the forest around them was on fire, their reindeer scattered in fright.

Nearly 200 km to the south, a man named Bryukhanov was plowing his land on a hillside above the Angara river. He had just sat down next to his plow to have some breakfast when suddenly he heard bangs that sounded like gunfire. His horse fell to its knees, and he saw flame shooting up into the sky to the north. Because there had been talk of war in the area, he thought the enemy was attacking. He then saw the trees in the forest bend over, as if they were hit by a hurricane. He grabbed hold of his plow with both hands, so the wind would not carry him away. While he held on, the wind swept by and picked up soil from his field. At the same time, he could see a wall of water being driven up the river.

These people, and many others in the region, had witnessed (and survived) an impact event. Later studies indicated that an asteroid or comet approached the area from the southeast. Hurtling through the atmosphere, it generated a brilliant fireball. As it neared the surface of the earth, sonic booms and a powerful shock wave radiated through the air. About 6 to 10 km above the ground, the object exploded, producing a radiant pulse of heat and a shock wave that slammed into the ground, followed almost immediately by an incredible blast of wind. The energy of the explosion was equivalent to 15 megatons of TNT, which is nearly 1000 times more powerful than the atomic bombs dropped in World War II.

The air blast produced by this tremendous explosion damaged a 2100-square-kilometer region of forest. Trees were blown down in an oriented fashion with their bases pointed towards the center of the blast. Upon close inspection, however, it was discovered that the tree-fall was not due simply to the final blast of the object, or it would have formed a simple circle around the point of the explosion. Scientists soon learned that the tree-fall pattern more closely resembled a butterfly with open wings, indicating that two shock waves affected the trees. The first shock wave was produced as the object hurtled through the atmosphere at supersonic speeds, while the second shock wave came from the explosion itself. The overlapping effects of these two shock waves produced the curious butterfly-shaped tree-fall pattern.

The effects of the impact were detected far beyond the Tunguska region. A pressure pulse traveled around the world twice, as indicated on barometric gauges as far away as Germany. Seismic waves generated by the blast were detected in Washington, Japan, and Java. Huge amounts of fine dust were thrown into the upper atmosphere. The evening sky in Europe and Asia was unusually bright for several nights, while the sunlight over North America was dimmed for weeks, illustrating that an impact in one region of the world can affect the entire planet.

What produced the Tunguska explosion? Was it a comet or an asteroid? A comet is believed to be so fragile that it would have exploded much higher in the atmosphere than the Tunguska object. An asteroid made of ordinary chondrite material, however, may have the right density to get within 10 km of the Earth's surface before blowing up. Based on its explosive energy of 15 megatons of TNT, the asteroid that produced the Tunguska explosion was calculated to be only about 30 m in diameter.

Meteor Crater, Arizona

The freshest, best-preserved impact crater on Earth is the one studied by Gilbert and Barringer in northern Arizona, where an iron asteroid about 50 m in diameter hit our planet, releasing the explosive equivalent of between 20 and 40 megatons of TNT. The impact carved out a 1.25-km-diameter crater as deep as the Washington Monument is tall (Fig. 9.5). Bedrock at ground zero vaporized, along with most of the

Figure 9.5 Meteor Crater, Arizona. (Image courtesy of Dan Durda)

impacting asteroid. Rock below and around the vaporized zone was ejected and overturned, burying the landscape out to a distance of approximately 2 km. Scientist David Roddy and his colleagues estimate that the explosion threw out 175 megatons of rock from the crater.

All of this happened during the last ice age, nearly 50,000 years ago. While the area is now a desert, at the time of the impact grasses and a pinon-juniper woodland covered the landscape. Instead of dry canyons, shallow streams cut the area as they flowed towards the Little Colorado River. Fossil evidence indicates mammoths, giant ground sloths, bison, and camels grazed on grass, and mastodons may have browsed on the bushes and trees.

When the meteorite hit and exploded, a shock wave radiated across the land, creating a very high speed wind called an *air blast.* In a recent study, David Kring (author of this chapter) determined that this air blast was one of the most important environmental consequences of the impact. If the asteroid hit with an energy of 40 megatons of TNT, the air blast would have produced winds in excess of 1000 kph that scoured the ground. The velocity diminished with distance, but hurricane-force winds raged at least as far as 40 km. These winds stripped away grass and soil, and destroyed pinon and juniper trees. Even beyond that distance, many trees were flattened or damaged.

When the air blast hit mammoths, ground sloths, and other animals, it severely damaged their lungs, hearts, brains, and other organs. All of the animals within a radius of 3 km were killed within minutes of the impact. Beyond the kill zone, many of the surviving animals suffered crippling injuries and severe lung damage.

In addition, animals died or received injuries when the blast wave hit them, picked them up, and then slammed them back onto the ground. This effect killed animals out to distances of 24 km from the impact site. The blast wave also picked up broken branches, rocks, and other types of missiles that could impale, lacerate, or traumatize animals. Stones and pebbles were accelerated to speeds of several hundred kilometers per hour. If a fusillade of these stones hit a standing mammoth it could blind, if not kill, the animal. The surviving animals also endured a rain of metal fragments from the small portion of the iron asteroid that was not vaporized. We know from the distribution of meteorites around Meteor Crater that iron projectiles rained down out to a distance of 9.5 km.

A pulse of heat energy compounded the devastating effects of the air blast. At Meteor Crater the thermal pulse may have scorched vegetation out to distances of about 10 km, though no evidence of this fire has yet been found. Perhaps the air blast immediately snuffed out any burning vegetation.

Though the Meteor Crater impact was a relatively small event, the direct and indirect effects of the blast wave were devastating to the local plants and animals at the time. This impact, however, was not large enough to have global or long-lasting consequences. Most of the damage was confined to a few hundred square kilometers, and plants and animals probably recolonized the area within decades. Millions of years before, the Earth was not so lucky.

Chicxulub, Mexico

One of the most dramatic events to affect Earth after the evolution of complex life occurred at the end of the Cretaceous Period, 65 million years ago, coinciding with one of the greatest mass extinctions in Earth's history. At that time at least 75 percent of the species on our planet, both in the seas and on the continents, were extinguished forever. The most famous of the vanquished are the dinosaurs. However, these giants were only a small fraction of the plants and animals that disappeared. In the oceans, more than 90 percent of the plankton was extinguished, which inevitably led to the almost total collapse of the entire food chain, thereby affecting every creature on Earth.

In the rock record, spanning the time when these species disappeared, there is a thin clay layer that is found at sites around the world. A team of scientists led by Luis Alvarez (a Nobel Prize–winning physicist) and his son Walter (a geologist) discovered that the clay layer contained a strikingly high concentration of iridium, an element that is much more common in meteorites than in Earth crustal rocks. They proposed that an impact was responsible for the mass extinction event. The discovery of high iridium concentrations in the clay layer at several places around the world suggested that the impact had been a large one.

These findings led to an extensive search for a large impact site 65 million years old. Seven researchers, including David Kring, finally located the impact on Mexico's Yucatan Peninsula (Fig. 9.6). It is a huge buried crater that was named Chicxulub, a Maya word that roughly translates as "tail of the devil." The crater, now buried beneath a kilometer-thick sequence of sediments, has been imaged using geophysical techniques that allow us to visualize underground structures. Deep boreholes, drilled through the overlying sediments, have yielded unmistakable samples of impact breccia. From these studies, we have found that a massive impact formed this huge crater. It appears to have a diameter of 145 to 180 km, which makes it one of the largest confirmed impact structures on Earth. Only Sudbury in Canada and the Vredefort structure in South Africa could potentially be larger.

The asteroid or comet that produced the Chicxulub crater was roughly 10 km in diameter. When an object that size hits Earth's surface, it causes a tremendous compression wave while transferring energy and momentum to the ground. This is similar to a large explosion, although the energy of the Chicxulub impact dwarfs anything in our experience. The energy of this impact was comparable to some 100 million megatons of TNT, 6 million times more energetic than the

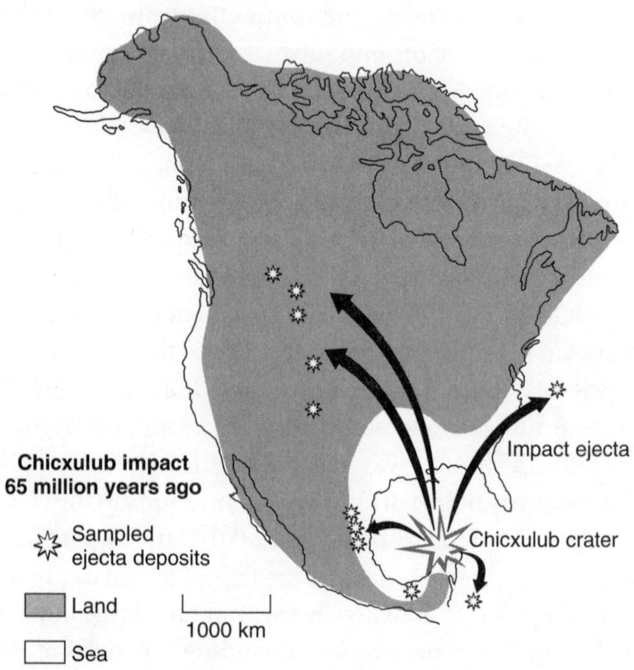

Figure 9.6 The Chicxulub impact crater and North America as it looked 65 million years ago.

1980 Mount Saint Helens eruption. The impact ejected rock from several kilometers beneath the surface of the Earth and carved out a bowl-shaped crater nearly 100 km in diameter. In addition, the shock of the impact produced magnitude-10 earthquakes—greater than any we have ever measured.

The initial bowl-shaped crater was very unstable, and its walls quickly collapsed along a series of faults that enlarged the final diameter to between 145 and 180 km. At the same time, the rock that had been compressed beneath the crater by the impact rebounded, producing a peak-ring structure in the crater's center. These dramatic changes, which rapidly transported huge volumes of rock over distances of tens of kilometers, occurred within only a few minutes. Because the impact site was in a shallow sea, water rushed in to fill the circular depression. Imagine the view as waterfalls tumbled over the rim of the crater and the fury of the water as it roared and hissed across the crater's burning floor.

Because ocean water filled and covered the crater, sediment on the bottom of the sea soon buried the impact scar. It is no longer visible today, even when standing directly over it. However, the structure has affected the circulation of groundwater on the Yucatan Peninsula, producing a ring of *cenotes* (springs). This ring of cenotes is nearly coincident with the rim of the Chicxulub structure and is the only visible feature on the surface to indicate that a huge crater lurks below.

The immense Chicxulub structure was the source of a tremendous amount of ejected debris that settled on all parts of the globe. The thickest portion of this debris blanketed North America and possibly South America. Near the impact crater the debris is tens to hundreds of meters thick, while as far away as Colorado (over 2000 km distance), the debris is still a centimeter thick. Some of the debris, including the gas from thc vaporized asteroid or comet, rose in a plume far above the Earth's atmosphere ultimately reaching halfway to the Moon.

Was this material enough to cause the mass-extinction event, and, if so, how did it cause the deaths of so many plants and animals?

At the moment, we do not know the complete answer to this question. To cause extinctions, an impact event must induce long-term changes in the physical or biological environment over such a large area that an organism could not escape by migrating or adapting. We know that some of the environmental changes produced by the impact affected only particular regions of the world, while other changes were global. Which of these changes or combination of changes caused the extinction of specific species is still unclear and that question is the focus of many current studies. Nonetheless, we can describe some of the environmental effects that occurred, even if we do not know how they affected specific organisms.

In the immediate vicinity of the impact, the blast and heat destroyed all signs of life. Forests were flattened to a distance of 500 to 1000 km. Because part of the crater was in a shallow sea, giant tsunamis radiated across the Gulf of Mexico. Geologic evidence of these waves crashing onto the coastline exist in Mexico and the United States. If the impact had occurred in a deep ocean basin, these waves might have been 4 to 5 km high and affected coastlines as far away as 10,000 km. Because the Chicxulub impact occurred in relatively shallow water, approximately 100 m deep, the waves were probably not nearly as large. One estimate suggests waves that hit the Texas coast were "only" 50 to 100 m high. Additional waves developed when millions of tons of ejected debris came crashing back down into the Gulf of Mexico and Caribbean basins. The powerful earthquakes that jolted the region immediately after the impact caused landslides along the coastline, generating more tsunamis.

Dust particles contained in the vapor plume rising above Earth's atmosphere blocked sunlight from reaching the surface around the entire globe. Current estimates suggest that the dust made it too dark to see for 1 to 6 months and too dark for photosynthesis for 2 months to 1 year, seriously disrupting marine and continental food chains. During this period, land surface temperatures decreased dramatically, possibly remaining below freezing in many areas. The dust eventually fell back to Earth, forming part of the global clay layer that marks the end of the Cretaceous period.

In addition to solid ejecta, toxic vapors were distributed worldwide. Carbonate and evaporite rocks covered much of the impact site. When the carbonate was vaporized it formed the greenhouse warming gas carbon dioxide. Vaporized evaporite released sulfur oxides, which react with water to produce sulfuric acid aerosols. These atmospheric particles significantly reduced the amount of sunlight reaching Earth's surface and thus enhanced the effects of ejected dust. The darkness and cooler temperatures produced by these particles likely lasted for months to years. Eventually, the sulfuric acid aerosols fell to the ground as acid rain.

When the solid ejecta reentered Earth's atmosphere, it shock-heated the air and dramatically altered its chemistry. The protective ozone layer was one casualty of this change. The formation of nitric acid rain added to the consequences of the sulfuric acid rain. One possible effect was defoliation of continental vegetation and even aquatic plants in shallow lakes or seas. Asphyxiation of animals by nitrous oxides and toxic poisoning by metals leached from the ground may also have occurred.

While some postimpact effects are still uncertain, one environmental catastrophe for which evidence exists is massive forest fires. Charcoal and soot have been found in sediments around the world. These fires were probably started when the impact debris came raining back down through the atmosphere. This debris quickly heated the air, producing a temperature spike so high that vegetation spontaneously ignited. These fires apparently consumed large quantities of vegetation and killed countless animals, either by burns or starvation.

Clearly, the Chicxulub impact caused severe environmental changes that could have led directly to the destruction of entire species. It is unlikely that any single process was responsible for all extinctions, but rather that they operated in concert. The time scales for these processes were also very different. The direct effects of winds and tsunamis lasted only a few hours, as did the temperature spike induced by falling ejecta. Fires, darkness, and cold temperatures probably lasted a few weeks to months and certainly less than a couple of years. The effects of acid rain and toxic substances may have lasted for several years, while greenhouse warming persisted for many centuries.

Other Mass Extinctions and Possible Links to Impacts

The Cretaceous event linked to the Chicxulub impact is one of five major episodes of extinction in the geologic record. The others occurred at the end of the Ordovician period (430 million years ago), during the Late Devonian period (350 million years ago), at the end of the Permian period (225 million years ago), and at the end of the Triassic period (200 million years ago). The mass extinction event at the end of the Permian was the most severe, claiming approximately 80 percent of the animal and plant species in the oceans. Some scientists speculate that all mass extinctions are related to impacts. However, none of these events is linked to an impact as strongly as the Cretaceous extinction is with the Chicxulub impact.

Some of the extinction boundaries do coincide with evidence for impact events. Shocked quartz, known to be produced by impacts, has been reported in sediments from near the end of the Triassic. Also, a layer of sediments dated near the end of the Devonian period contains impact-melt spherules. It is unclear, however, whether these impact products come from the exact time of the extinction events. Nor is it clear yet whether the impacts that produced the shocked quartz and impact-melt spherules were large enough to have caused mass extinctions. The 100-km-diameter Manicouagan crater in Canada was once thought to be linked to the terminal Triassic event, but recent analyses indicate the crater dates from about 12 million years before the extinction.

One of the most intriguing coincidences between impact debris and species extinction involves a relatively recent series of events during the late Eocene, about 34 million years ago. These events, spanning a period of 10 million years, combined to form the largest group of extinctions since the dinosaurs' demise. A blanket (and possibly multiple blankets) of impact-melt spherules appear in sediments deposited during this interval of time. This evidence indicates there was at least one impact near the time of the late Eocene extinctions, probably somewhere in North America.

Scientist Wylie Poag and his colleagues have recently discovered a large crater beneath sediments in the Chesapeake Bay. Although it is completely buried, about 50 wells have been drilled into the crater and its surrounding ejecta blanket. Analysis of rocks from the wells and subsurface images produced by geophysical techniques have revealed a crater diameter of as much as 90 km. The ejecta blanket extends over an area greater than 8000 square kilometers in southeastern Virginia and on the adjacent ocean floor. Both the crater and the impact melt spherules are 35 million years old. A second feature of the same age is the Popigai crater in Siberia. Its diameter of 100 km is even larger than the Chesapeake Bay crater. In fact, Popigai is the fourth or fifth largest crater known on Earth.

The Chesapeake Bay and Popigai impact events occurred only a few hundred thousand years apart, or possibly less. Consequently, it appears Earth was rocked with a one-two punch—a potentially devastating combination. However, some paleontologists argue that the extinctions they have identified occurred either long before or long after the impacts, which leads them to conclude that the impacts are not related to the extinctions. Nonetheless, these two impacts were so large that they clearly disrupted the environment around the entire world.

At the moment, the link between the late Eocene impacts and extinctions is tenuous, and further studies are obviously needed. (This is the same situation that once existed in studies of the Cretaceous extinction, but detailed research finally demonstrated that the extinction and impact happened at essentially the same time.) The coincidence or near coincidence of these phenomena in the late Eocene raises two interesting questions. If impact events producing 90- and 100-km-diameter craters did not cause or at least markedly contribute to the extinctions, what natural process could have had a more dramatic and lethal effect? Global cooling and changes in oceanic circulation are the favorite alternatives among those who dismiss the impact-cratering hypothesis. If we assume that the impact events are not related to the extinctions, then this raises a second question: What size impact does cause extinctions? In broader terms, we might also ask what sizes of impact events cause regional disasters versus global disasters and if the effects depend on whether the impact occurs in a sea or on a continent. We must answer these questions in order to interpret accurately events that occurred long ago in Earth's history, and properly evaluate the threat of asteroids and comets that *will* inevitably hit the Earth in the future.

PART FIVE

Assessing the Risk

EPILOGUE

The Threat of Future Impacts on Earth

Recent Near Misses

The March 12, 1998, front-page headline of the *New York Times* read, "Asteroid Expected to Make A Pass Close to Earth in 2028." The first sentence of the article went on to say "there is a very slight possibility that it might hit Earth" and then the article explained that if the asteroid hit Earth it "would have devastating global effects, including tidal waves, continent-size fires and an eruption of dust that could cause global cooling and long-term disruption of agriculture." It all sounded like media hype for the summer's twin disaster movies, *Deep Impact* and *Armageddon*—but this was real.

Not surprisingly, the story led to near panic. National news sources of all types scrambled to get experts on the air or in print with explanations—How bad would it be? Where would it hit? Do we have time to do anything about it?

The threat was a newly discovered asteroid called 1997 XF11. It was first seen by Jim Scotti with the University of Arizona's Spacewatch telescope, an instrument specially designed to detect objects in near-Earth space. He discovered the asteroid in December 1997, months before the news reports flashed their warnings. His initial observations did not provide enough data for astronomers to calculate the asteroid's orbit. Other astronomers searched for the object and it was reacquired by a Japanese team before the end of the month and again by a Texas observer in early March. These three sets of observations permitted calculation of a preliminary orbit. This calculation showed that the asteroid would pass close to Earth, and with so few observations over such a short period of time, astronomers were unable to rule out the chance that it might actually hit our planet. This uncertainty made headlines and led to the explosion of concern.

Immediately, astronomers went to work checking their photographs to see if the object might have been captured previously. If so, that would increase the number of observations and lengthen the time between first and last positions, reducing the uncertainty in the calculated orbit considerably. Fortunately, within hours of the initial report, observers discovered older photographs of the asteroid and determined that the object would miss Earth by several hundred thousand kilometers. Nonetheless, with all of the news coverage, the modern-day threat of impact events became real to many people.

Several other near misses have been reported in the past few years. One of the most memorable for coauthor David Kring occurred in December 1994. At the time, Kring was at Meteor Crater, being interviewed for a science documentary about impacts. One of the last things he explained was how the odds of impacts really work. Scientists are always saying that objects of various sizes hit the Earth at average rates of once per 10,000 years, once per 100,000 years, once per million years, and so on. These numbers are statistical; that is, we cannot use them to predict that an object of a certain size will hit at a particular time in the future. Rather, as he explained in the interview, because these are statistical averages it is also possible for an asteroid of any size to hit tomorrow.

Much to Kring's surprise, while driving back to Tucson the next day, he heard a report on the radio that said an asteroid had just passed perilously close to the Earth. While he was correct to warn that an impact could happen at any time, it was a relief to learn that that time would not be now. As it turns out, Jim Scotti also discovered this asteroid, which was designated 1994 XM1. It is the closest-approaching asteroid detected thus far by astronomers. 1994 XM1 passed within 112,000 km of Earth—less than one-third the distance to the Moon. This asteroid was the size of a house, or slightly smaller than the object which exploded over Tunguska.

1994 XM1 is certainly not the only near-Earth object. Scotti estimates that, on the average, there are about 40 house-sized objects closer than the Moon at any one time, and about 100 pass within that distance each day. In other words, several thousand metric tons of rocks are likely to be cruising through the Earth's neighborhood at this very moment, and we have no idea where any of them are.

Larger asteroids also orbit in the vicinity of Earth, although they are less abundant than the house-sized objects. One of these came uncomfortably close in May 1996. This particular asteroid was 500 m across and whizzed by only 446,000 km from Earth. We know about the object because two students, Tim Spahr and Carl Hergenrother, spotted it through a telescope in southern Arizona. The asteroid came hurtling down from above the plane of the solar system—a part of the sky that most asteroid surveys do not examine. The students found the asteroid just days before its closest approach, which is far too little time for us to have done anything to prevent a collision. How many more menacing rocks lurk in our part of space?

These new observations are changing the way we think of Earth. Where once we pictured our planet as a beautiful but isolated blue ball in space, we now realize that it is part of a solar system which is a bit too crowded for comfort. Space is not actually getting more crowded, it is just that we were once blissfully ignorant of the hazards in the vicinity of our planet. As new telescope searches and more accurate computer models are devoted to the search for near-Earth objects, a better and more complete picture of our space environment will continue to unfold.

Impact Rates

The introduction to this book related the sizes of impacting objects to their effects on Earth, and estimated how often, on average, they hit our planet. Where did these estimates come from?

We use three sets of data to estimate the frequency of impacts on Earth. The first is the terrestrial cratering record. As discussed in Chap. 9, about 150 impact craters have been found thus far, some recently formed and others dating back more than 2 billion years. However, the record preserved on Earth is far from complete. Erosion, deposition, and tectonic activity have destroyed most of our planet's craters. These processes are particularly effective at getting rid of old craters and small craters. Also, it is difficult to find impact craters on the seafloor, which represents about 70 percent of Earth's surface. Nonetheless, Richard Grieve and Gene Shoemaker have calculated that impacts form a crater 20 km in diameter or larger about once every million years.

Because the cratering record on the Earth is so incomplete, scientists use crater counts from the Moon to infer the impact rate on Earth. In a recent analysis of the lunar data, Gerhard Neukum and Boris Ivanov estimated that a 1-km crater (the size of Meteor Crater) is produced every 2000 years, a 10-km crater every 260,000 years, and a 100-km crater every 27 million years.

The third data set comes from telescope searches for asteroids and comets in Earth-crossing orbits. The current inventory is far from complete, but scientists estimate that there may be 2000 Earth-crossing objects larger than 1 km, as many as 8000 such objects larger than 500 m, and perhaps a million larger than 50 m (the size object that produced Meteor Crater).

Asteroid and Comet Impacts—What the Future Holds

In 1992, at the request of Congress, NASA evaluated the hazards of impacting objects and recommended a strategy for detecting them. That report, which is called the Spaceguard Survey, divided near-Earth objects into three categories, according to their size and the magnitude of the threat they pose for Earth.

The first category includes impactors up to 100 m in diameter. An object of this size, hitting the Earth with a typical asteroid velocity of 72,000 kph, has the kinetic energy of 50 to 100 megatons of TNT—the explosive energy of our largest thermonuclear weapons. Depending on the size and strength of the objects, they may explode in the atmosphere or smash into the ground, with devastating local consequences in either case. An asteroid around 30 m in diameter flattened the forests of Tunguska in 1908. As previously discussed, impacts of this magnitude occur on the average every few hundred years.

Impactors in the second category are 100 m to 1 km in diameter. The largest of these objects hit on average once every few hundred thousand years. While we do not understand impact events this large in detail, we believe they may have enough energy to produce global consequences.

The third category includes impactors 1 to 5 km in diameter, which hit the Earth about once in tens of millions of years. These are definitely over the threshold for producing global effects, and may be responsible for species extinctions. The severity of global effects would depend on the object and the target material, but any impact in this size range could challenge our very survival.

What might happen to our world if one of these large asteroids or comets should hit? If, for example, the impact is in an ocean as with the Chicxulub impact, one of the consequences will be tsunamis. These powerful waves will wash over the surrounding coastlines and wreak havoc. Since a large fraction of the world's population lives in coastal cities, the repercussions of tsunamis are very important. To illustrate their potential effects, consider two hypothetical impacts in the Atlantic Ocean.

Imagine a 100-m object hitting about 200 km east of Cape Hatteras, North Carolina. The explosive energy of 50 megatons of TNT would produce a massive wave that would crash over the coastline. As it approached shallow water, the wave would grow to over 100 m high and would sweep approximately 20 km inland.

Now imagine a 1-km object with the explosive energy of nearly 50,000 megatons of TNT impacting the Atlantic Ocean 650 km east of New York City. A 100-m-tall wave would race away from ground zero, but near the shore it would grow to nearly 1000 m in height. This incredible wall of water would destroy New York and severely damage many other East Coast cities, and would wash inland for hundreds of kilometers.

These are relatively large impact events, but they are the type that occur on average every thousand to hundreds of thousands of years. Because water covers two-thirds of the Earth, the majority of impacts like these will occur at sea, spawning disastrous tsunamis.

Of course, impacts do occur on land, as well. Recall that Meteor Crater was produced when an iron asteroid, about 50 m in diameter, collided with the Earth with the explosive equivalent of 20 to 40 megatons of TNT, carving out a 1.25-km-diameter crater as deep as the Washington Monument is tall. What would happen if an asteroid of this size were to hit Earth again?

While the Meteor Crater impact event occurred before humans had settled in the region, it is in an area that now contains several towns and a small city. If the impact occurred today in the same location, it would undoubtedly produce a multitude of human casualties. Of course, when compared to other areas of the world, this region of northern Arizona is relatively uninhabited. In more densely populated areas, the effects of impacts this size could be even more catastrophic. Tens of thousands to millions of people could be killed if an event of this size occurred unexpectedly over a modern city.

For example, if you superimpose the effects of the Meteor Crater impact event on a metropolitan area the size of Kansas City, the results are obvious—immense destruction (Fig. E.1). The Kansas City metropolitan area is roughly the same size as the region damaged by the Meteor Crater event. An air blast within this region would shatter windows, cause failure and buckling of corrugated steel or aluminum paneling, shatter cinderblock and brick walls, and blow out wood siding. The ruined city would be further devastated by fires from the heat pulse and broken gas lines. The computer-simulated destruction of our world that Hollywood has portrayed is only a hint of the possible devastation.

Do we need to worry about this scale of impact hazard? The answer may well be yes. Current estimates suggest that impacts this size occur every few thousand years globally. This means they are relatively frequent, even on a time scale that humans find meaningful. Furthermore, as the population of our world increases and we occupy larger

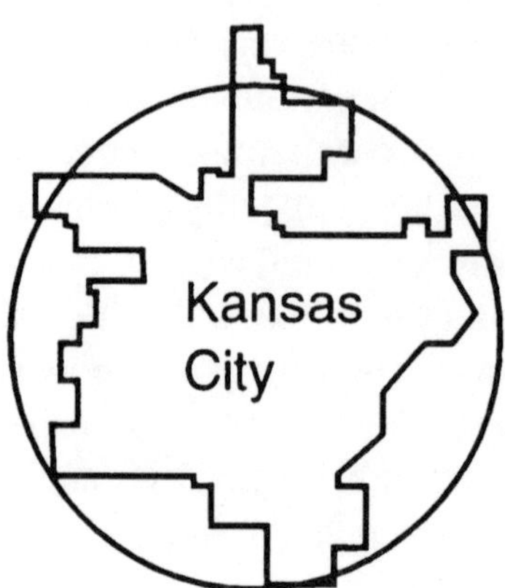

Figure E.1 Map of Kansas City, Missouri, with a 40-km-diameter circle–the area of severe damage from a Meteor Crater–sized impact.

fractions of the Earth's surface, the chances of this type of event occurring near a city rather than in a desolate area will increase.

Deflecting Hazardous Asteroids and Comets

What are our options if an asteroid or comet is on a collision course with Earth? We basically have two choices—we can either try to destroy the object or we can try to deflect it to miss the Earth.

It is very difficult to destroy an asteroid, particularly one the size of a mountain or larger. Scientists in the nuclear weapons industry have suggested powerful explosive devices that might fragment the asteroid. Other analysts who have examined the issue are not convinced the object could be completely destroyed. They also worry about the political instability that systems with such destructive capacity would produce in the world.

Putting aside the political question, it is important to understand why complete destruction of an asteroid may be extremely difficult. When a mass of rock is broken, the largest piece is usually about half the size of the original object. Consequently, if one could break a 1-km-diameter asteroid, the largest fragment heading for Earth might still be about 500 m in diameter. We would need a way to fragment the object so completely that the largest pieces were less than 10 m across. Ten-meter-size objects hitting the Earth could still cause considerable damage, but the event would be globally survivable.

Since we do not have any experience in blowing up large asteroids, we cannot calculate the amount of energy it would require with any certainty. We know there is a range of physical strengths in asteroids; thus, there is a range in the amount of explosives needed to destroy them. Tom Ahrens and Alan Harris have estimated that the energy needed to fragment a 1-km asteroid is from 1 to 3 megatons of TNT. Destroying larger bodies may require energies on the order of 3000 megatons of TNT. This is far beyond anything in our nuclear arsenals.

The location of any explosion is also critical. For maximum destruction it must be placed near the center of the asteroid. Blockbuster science-fiction movies like *Armageddon* notwithstanding, we do not yet know how to drill to the center of a 10-km asteroid. For these reasons, catastrophic fragmentation is not a likely option for dealing with a major asteroid threat in the near future.

If we cannot destroy hazardous objects, can we deflect them? Several proposed scenarios depend on the size of the object and the amount of time before collision.

The earlier we detect an object, and the earlier it is deflected, the better. A small orbital change can cause an asteroid to miss Earth if that change is made years in advance. If the object is near Earth, much more energy is needed to change the orbit to a safe one. If the object is too close before being discovered, it may be impossible to deflect it in time to miss Earth.

One way to deflect an asteroid is to crash a rocket into it. This will produce a crater and eject debris into space, pushing the asteroid into a slightly different orbit. Calculations suggest this is a feasible method for objects 100 m across or less, but that it is useless against larger bodies.

For slightly larger asteroids, up to about 1 km in diameter, one could eject material over an extended period of time. A facility on the surface of the asteroid could mine rock and toss it into space, each toss pushing the asteroid slightly in the opposite direction and changing its orbit. To deflect a 1-km asteroid well in advance of a collision with Earth, the facility might need to eject 10,000 tons of rock, with a velocity of 100 kph, over a period of 10 years—a significant but not impossible space mining project.

We might also use devices that operate above the surface of an asteroid, heating the surface and jettisoning material off into space (Fig. E.2). For example, one could detonate a nuclear device adjacent to the asteroid. Energetic neutrons and gamma rays would radiate from the point of explosion and deposit energy on one side of the asteroid, which could cause material to expand and break off, producing a momentum pulse in the opposite direction and a change in orbit.

Heating one side of the asteroid with pulsed lasers, microwaves, or solar collectors might also work. If they deposit enough energy, these techniques will vaporize material which will be jetted into space. Like a rocket engine, these jets will push the asteroid in the opposite direction and change its orbit.

All of these scenarios rely on calculations. There are no efforts in place to design or test any of these systems, and the technical difficulties would be enormous. For example, the solar collector system needed to superheat an asteroid's surface might well involve a fleet of mirrors, each 10 km in diameter, orbiting with the asteroid for over a decade. Because of the costs involved, and because some of the proposed tech-

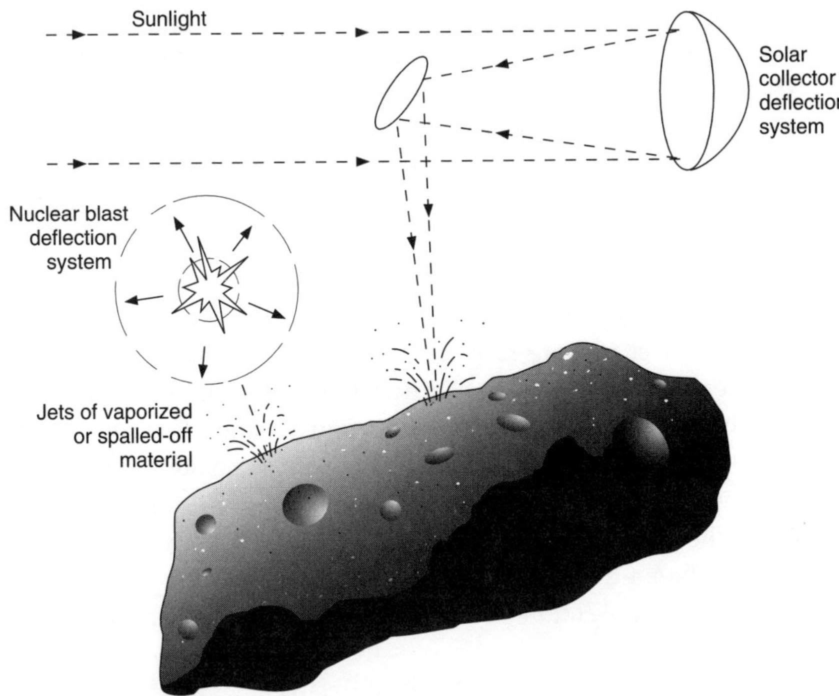

Figure E.2 Techniques for deflecting an asteroid.

nologies could be used as weapons, it is not likely that any of the systems will be built until a real threat is discovered.

There are other important issues to address in the meantime. First, we can find the asteroids in Earth-crossing orbits and determine which ones might hit Earth in the near future. This is a formidable task that will require dedicated attention by several astronomical observatories. The cost should be relatively modest, however. Current estimates suggest that we can identify 90 percent of the near-Earth objects larger than 1 km in 10 years. This effort will identify those asteroids large enough to destroy our civilization. The estimated annual cost for this global early warning is around $4 million—much less than the production costs of those popular Hollywood impact movies.

Do We Have a Future?

We have reached a stage in our evolution where we can foresee our own destruction. In what many people have described as a global awakening, we now realize there are processes with global effects, and that some of these may be so severe that they threaten the survival of our species. Many of these potentially devastating effects are created by humans—nuclear war, global warming, destruction of the protective ozone layer, and species depletion that threatens the biologic diversity of our planet. In addition, we now understand that comets or asteroids, over which we have no control, have in the past and can again extinguish most of the life on our planet.

This brings us to two important points. First, as so many scientists have warned before, it is not a question of *whether* a major impact event will occur again—rather, it is a question of *when.* With considerable certainty, we know that Earth will be hit again by large asteroids or comets sometime in the future. Second, we know that even smaller, more frequent, impacts can threaten our lives and our civilization.

Does that mean we are in imminent danger? Probably not, because the average time span between globally destructive events is far greater than a human lifetime. However, this is a statistical average that provides only limited comfort—an impact of any size can occur at any time.

Now that we have arrived at this precipice of awareness, what do we do?

It is imperative that we seriously study the impact hazard and make realistic plans to deal with it. We should support a dedicated astronomical survey to locate the near-Earth asteroids and comets and calculate their orbits, in order to determine which are most likely to collide with Earth. We should carefully study the effects of past impacts, so that we can better predict the consequences of future hazards. Finally, we must devise workable defenses against a comet or asteroid on a collision course with our planet. We do not need to construct these systems, but they should be sufficiently well understood to allow a rapid and effective reaction when an impending collision is detected.

The actions we take could save the life we know on Earth today or our descendants and their world in the future.

INDEX

Aerogel, 59
Ahrens, Tom, 175
Albedo (of asteroids), 127
Allan Hills 81005 meteorite:
  age of, 104
  lunar origin of, 100
Allan Hills 84001 meteorite:
  history of, 109
  life forms on Mars and, 112–113
  Martian origin of, 109
Alvarez, Luis, 159
Alvarez, Walter, 159
Amor asteroids, 130
Andromedid meteor shower:
  comet Biela and, 73–75
  described, 73–74
Apollo asteroids, 130
Archaeoastronomy, 40–41
Arend-Roland (comet). *See* Comet Arend-Roland
Aristotle, 33
Armageddon (motion picture), 169, 175
Asher, David, 41
Asteroid belt, 122–123, 128–129
Asteroid collisions, 128–129
Asteroid families:
  Eos family, 129
  Kiyotsugu Hirayama and, 128
  Koronis family, 129
  Themis family, 129
Asteroids:
  Amor asteroids, 130
  Apollo asteroids, 130
  Braille, 58
  Ceres, 119–122, 127
  compared with comets, 121
  C-type, 127
  dark C type, 127
  Eos family, 129
  Eros, 143–145
  infrared observation of, 126–127
  Juno, 122
  lagrangian points and, 124
  measuring size and shape of, 124–127
  Mimistrobell, 60
  M-type, 127
  naming of, 122
  near-Earth asteroids, 130
  1994 XMI, 170–171
  1997 XF11, 169–170

Asteroids (*Cont.*):
numbering of, 122
number of, 122–123
occultation of, 125
origins of, 121–122
Pallas, 122, 125, 126
reflectance spectra of, 127
Rodari, 60
size of, 122–123
S-type, 127
Trojan asteroids, 124
Vesta, 122, 125
Astronomical units, 120
Astrophotography, 25, 42
A-type asteroids, 127
Azzurra, 141

Barnard, Edward Emerson, 42
Barringer, D. M., 153
Bath Furnace meteorite fall, 91–92
Bayeux Tapestry, 34
Beljawsky (comet). *See* Comet Beljawsky
Bennett (comet). *See* Comet Bennett
Bessarion B (lunar impact crater), 153–154
Bethlehem Star, 31–32
Biela (comet). *See* Comet Biela
Biela, Wilhelm von, 72
Blennert, John, 95
Bode, Johann, 120
Bonaparte, Napoleon, 38
Bondone, Giotto de, 32
Borrelly (comet). *See* Comet Borrelly
Bradley, Francis, 73
Brahe, Tycho, 34–35
Braille (asteroid), 57
Breccia, 101
Brooks (comet). *See* Comet Brooks
Brownlee, Don, 77–78

Cape York meteorite, 82
Ceres (asteroid), 119–122
Charon, 21
Chassigny meteorite:
described, 98, 99, 108–109
Martian origin of, 98, 99, 108–109
Chesapeake Bay impact crater, 164
Chicxulub impact event:
described, 4, 159–163
effects of, 4, 161–163
Chladni, Ernst, 119
Chondrites, 83, 89
Chondrules, 83
Christmas Star, 32
Close encounters (comets), 47–61
Clube, Victor, 41
Coggia (comet). *See* Comet Coggia
Collision threats (asteroids):
asteroid 1994 XM1, 170–171
asteroid 1997 XF11, 169–170
Collision threats (comets):
comet Encke, 50–51
comet Swift-Tuttle, 50
Comet Arend-Roland, 27
Comet Beljawsky, 27

Comet Bennett, 26
Comet Biela:
  Andromedid meteor shower and, 72–74
  breakup of, 47, 73
Comet Borrelly, 57
Comet Brooks, 27
Comet close encounters, 47–61
Comet Coggia, 27
Comet d'Arrest, 57–58
Comet de Kock-Paraskevopoulos, 27
Comet Donati, 27
Comet dust:
  collecting, 77–79
  meteor showers and, 65–66
  spectral analysis of, 76–79
Comet Encke:
  as collision threat, 50–51
  Comet Nucleus Tour (CONTOUR) space probe and, 57–58
  discovery of periodic nature, 37–38
  Taurid meteor shower and, 50
  Tunguska impact event and, 51
Comet Hale-Bopp:
  brilliance of, 24
  described, 26, 40, 46
  Heaven's Gate cult tragedy and, 40
  Hubble telescope observations of, 16
  SOHO spacecraft observations of, 46
Comet Halley:
  coma of, 16–17
  composition of, 44
  debris from, 48–49
  described, 26–28, 33–34, 38–39
  events attributed to, 33–34, 38
  history of, 33–34, 38–39
  nucleus of, 15, 19
  orbit of, 20
  Orionid meteor shower and, 75
  periodic nature of, 35–37
  photograph of, 37
  spacecraft observations of, 15, 43–44
Comet Hind, 27
Comet hunting, 42–43
Comet Hyakutake:
  described, 26, 29–31, 40
  photograph of, 30
Comet Ikeya, 43
Comet Ikeya-Seki:
  brilliance of, 24
  described, 26, 43, 48
Comet Klinkerfues, 27
Comet Kohoutek:
  coma of, 16
  depictions of, 39
  described, 26, 39
  ion cloud, 17
Comet Mrkos, 26
Comet names, 41–43
Comet Nucleus Tour (CONTOUR) space probe:
  comet d'Arrest and, 58
  comet Encke and, 58

Comet Nucleus Tour (CONTOUR) space probe (*Cont.*):
  comet Schwassmann-Wachmann 3 and, 58
Cometomania, 35
Comet phobia, 32–35
Comet photography, 25
Comet Pons, 28
Comet probes:
  Comet Nucleus Tour (CONTOUR), 58
  Deep Impact, 59
  Deep Space 1, 58
  Rosetta, 60
  Stardust 1, 58–59
Comets:
  Christmas Star and, 32
  coma of, 15
  compared with asteroids, 121
  debris from, 48–49
  defenses against, 175, 178
  defined, 11
  early history of, 32–35
  head of, 17
  ion cloud, 17
  list of, 26–28
  long-period comets, 20
  nucleus dynamics, 17–19
  nucleus of, 15–17
  origins of, 11–15, 20
  parts of, 15–19
  periodicity theory and, 22–23
  short-period comets, 20
  solar wind and, 19
  spectra of, 17
  Star of Bethlehem and, 31–32
  tails of, 19
Comet Schwassmann-Wachmann, 58
Comet Seki-Lines, 26
Comet Shoemaker-Levy 9:
  collision with Jupiter, 22, 51–56, 152
  history of, 51–52
Comet Skjellerup-Maristany, 27
Comet SOHO-6, 48
Comet Swift-Tuttle:
  as collision threat, 50
  Perseid meteor shower and, 71
Comet Tebbutt, 27
Comet Tempel 1, 59
Comet Temple-Tuttle, 72
Comet Thatcher 1861 I, 75
Comet Tuttle, 75
Comet viewing, 23–25
Comet Wells, 27
Comet West, 26, 39, 47–48
Comet White-Ortiz-Bolelli, 26
Comet Wild 2:
  described, 58
  Stardust space probe and, 58–59, 79
Comet Wilson-Hubbard, 26
Comet Wirtanen:
  described, 60
  Rosetta space probe and, 60–61
Consolmagno, Guy, 97
CONTOUR. *See* Comet Nucleus Tour space probe
Coon Mountain impact crater. *See* Meteor Crater impact event
Cosmic Background Explorer (COBE) spacecraft, 132

Cryosphere (Martian), 110
C-type asteroids, 127

d'Arrest (comet). *See* Comet d'Arrest
d'Arrest, Heinrich, 74
Daylight Comet:
  described, 27
  See also Great January Comet of 1910
Death star, 23
Debris (comets), 48–49
Deep Impact (motion picture), 169
Deep Impact space probe mission, 59
Deep Space 1 comet probe, 58
Deimos, 134–135
Delta Aquarid meteor shower, 69
Descartes, Rene, 20
Donahue, Robert, 93
Donahue, Wanda, 93
Donati (comet). *See* Comet Donati
Dust (interplanetary):
  collecting, 77–79
  meteor showers and, 65–66
  spectral analysis of, 76–77

Earth:
  early history, 14
  See also Impact craters; Impact events
Eclipse Comet, 27
Edgar Wilson Award, 43
EET 79001 meteorite, 103
Elliot, Andrew, 66
Encke (comet). *See* Comet Encke
Encke, Johann Franz, 37–38
Eos asteroid family, 129
Eros (asteroid), 142–145
Eta Aquarid meteor shower:
  comet Halley and, 75
  described, 69
Extinctions. *See* Mass extinctions

Fireballs, 93–94
Fusion crust (meteorite), 88

Galactic plane, 23
Galaxy, 23
Galileo spacecraft:
  comet Shoemaker-Levy 9–Jupiter collision and, 52, 54–55
  Deimos and, 134–135
  Gaspra and, 135–138
  Ida and, 138–142
  Phobos and, 134–135
Gaspra (asteroid):
  Galileo spacecraft and, 135–136
  properties of, 136–138
Gegenschein, 66
Geminid meteor shower:
  asteroid 3200 Phaeton and, 75
  described, 69
Gilbert, G. K., 153
Giotto spacecraft, 15
Great Comet, 27
Great Comet of 1811:
  coma of, 17
  events attributed to, 38
Great Comet of 1830, 28
Great Comet of 1831, 28
Great Comet of 1882, 47

Great January Comet of 1910:
described, 38
See also Daylight comet
Great March Comet, 27
Great September Comet, 27
Great Southern Comet, 27
Grieve, Richard, 172

Hale-Bopp (comet). *See* Comet Hale-Bopp
Halley (comet). *See* Comet Halley
Halley, Edmund:
Halley's comet and, 35–37
Isaac Newton and, 35–36
Harold, King, 34
Harris, Alan, 175
Heaven's Gate cult tragedy, 40
Hergenrother, Carl, 171
Herrick, Edward, 73
Hevelius, Johann, 35
Hirayama, Kiyotsugu, 128–129
Hoba meteorite, 82
Hsin, Chou, 33
Hubble Space Telescope:
asteroids and, 125
comet Hale-Bopp and, 16
comet Hyakutake and, 31
comet Shoemaker-Levy 9–Jupiter collision and, 52, 53, 55
Humboldt, Alexander von, 68
Hyakutake (comet). *See* Comet Hyakutake
Hyakutake, Yuji, 29

Ida (asteroid):
Azzurra (crater) and, 141
Dactyl (moon) and, 142
Ida (asteroid) (*Cont.*):
Galileo spacecraft and, 138
properties of, 138–142
Ikeya (comet). *See* Comet Ikeya
Ikeya, Kaoru, 43, 48
Ikeya-Seki (comet). *See* Comet Ikeya-Seki
Impact craters:
Chesapeake Bay crater, 164
Chicxulub crater, 4, 159–163
double craters, 153–154
East Clearwater Lake crater, 154
Gusev crater, 154
Haviland crater, 153
Kamensk crater, 154
Meteor Crater, 153, 157–159, 173
Popigai crater, 164–165
Ries crater, 154
Sobolev crater, 153
Steinheim crater, 154
Sudbury crater, 153
Tunguska crater, 155–157, 172
Vredefort crater, 153
West Clearwater Lake crater, 154
Impact events:
Chicxulub explosion, 4, 159–163
Meteor Crater explosion, 153, 157–159, 173–174
Tunguska explosion, 155–157, 172
See also Impact craters
Impact rates, 171–172
Infrared Astronomical Satellite, 127

Innisfree meteorite, 88–89
International Star Registry, 42
Interstellar dust, 83–84
Ion cloud (comets), 17
IRAS. *See* Infrared Astronomical Satellite
Ivanov, Boris, 172

January Comet of 1910, 24
Jilin meteorite fall, 92
Jull, Tim, 96
Juno (asteroid), 122
Jupiter:
- asteroid belt and, 123–124
- birth of, 12
- comet Shoemaker-Levy 9 and, 22, 51–56, 152
- lagrangian points and, 124

Kant, Immanuel, 20
Kegler, Ignatius, 50
Keppler, Johannes, 35, 36, 120
Kessuh, 82
Kinetic energy, 6
Kirkwood, Daniel, 123–124
Kirkwood gaps, 123–124, 130
Klinkerfues (comet). *See* Comet Klinkerfues
Klinkerfues, Wilhelm, 75
Knapp, Michelle, 93
Kohoutek (comet). *See* Comet Kohoutek
Koronis asteroid family, 129, 142
Kreutz, Heinrich, 48
Kreutz sungrazing comets, 48
Kriegh, Jim, 95
Kring, David, 94–96, 158, 159
Kuiper, Gerard, 20
Kuiper belt, 20, 21

Lagrange, Joseph-Louis, 124
Lagrangian points, 124
Leonid meteor shower:
- comet Temple-Tuttle and, 72
- described, 66–67, 69, 75

Levy, David, 51
Long-period comets, 20
Lost City meteorite, 88
Lunar impact craters, 149
Lunar meteorites:
- age of, 104
- Allan Hills 81005 meteorite, 100, 104
- evidence pointing to moon as source, 101
- history of, 103–107
- list of, 101

Lyrid meteor shower:
- comet Thatcher 1861 I and, 75
- described, 69

Mariner 9 spacecraft:
- Mars and, 133–134
- Martian moons and, 134

Mars:
- Allan Hills 84001 meteorite and, 109, 112–113
- Chassigny meteorite and, 98, 99, 108–109
- Deimos and, 134–135
- impact craters on, 151, 152
- Lewis Cliff 88516 meteorite and, 109
- Martian meteorites and, 99–106, 108–110

Mars (*Cont.*):
  Nakhla meteorites and, 99, 108
  Phobos and, 134–135
  Shergotty meteorite and, 99
  SNC meteorites and, 101–103
  Viking spacecraft and, 111
  water on, 110–111
  Zagami meteorite and, 100
  See also Martian meteorites
Marsden, Brian, 50
Martian meteorites:
  age of, 105
  Allan Hills 84001 meteorite, 109, 112–113
  Chassigny meteorite, 98, 99, 108–109
  color of, 103
  evidence pointing to Mars as source, 101–103
  history of, 103–106, 108–110
  Lewis Cliff 88516 meteorite, 109
  list of, 100
  Nakhla meteorites, 99, 108
  Shergotty meteorite, 99
  Zagami meteorite, 100
Mass extinctions:
  Chicxulub impact event and, 159–163
  Cretaceous extinction, 159–163, 165
  impact events and, 163–164
McKay, David, 112
Mercury:
  impact craters on, 150–151
  meteorites from, 114
Messier, Charles, 42
Meteor Crater impact event, 153, 157–159, 173–174
Meteor dust, 65–66
Meteorite falls:
  examples, 91–93
  frequency, 90–91
  size, 90–91
Meteorite hunting:
  in Antarctica, 96–97
  in deserts, 94–96
Meteorites:
  age of, 89–90, 104, 105
  Allan Hills 81005 meteorite, 100, 104
  Allan Hills 84001 meteorite, 109, 112–113
  Bath Furnace meteorites, 91–92
  Cape York meteorite, 82
  Chassigny meteorite, 98, 99, 108–109
  composition of, 86–88
  described, 82
  disposition of, 98
  Hoba meteorite, 82
  identifying, 88–89
  Innisfree meteorite, 88–89
  Jilin meteorites, 92
  Lewis Cliff 88516 meteorite, 109
  Lost City meteorite, 88
  Nakhla meteorites, 99, 108
  Shergotty meteorite, 99
  Sikhote-Alin meteorites, 92
  Zagami meteorite, 100
  See also Lunar meteorites; Martian meteorites
Meteor showers:
  Delta Aquarids, 69
  Eta Aquarids, 69

Meteor showers (*Cont.*):
  Geminids, 69
  Leonids, 69
  Lyrids, 69
  Orionids, 69
  Perseids, 69
  Quadrantids, 69
  Ursids, 69
Milky Way, 23
Mimistrobell (asteroid), 60
Molecular clouds, 83
Monrad, Ingrid, 95
Montaigne, 72
Moon:
  impact craters on, 149
  lunar meteorites, 99–107
  water on, 13
Mrkos (comet). *See* Comet Mrkos
M-type asteroids, 127

Naked-eye comets, 26–28
Nakhla meteorites, 99
Napier, Bill, 41
Near-Earth Asteroid Rendezvous (NEAR) spacecraft:
  Eros and, 143–144
  Mathilde and, 143–144
Near-Earth objects:
  Amor asteroids, 130
  Apollo asteroids, 130
  exploration of, 145–148
  frequency of, 169–171
  impact rate of, 171–172
  mineral resources on, 147–148
  1994 XMI, 170–171
  1997 XF11, 169–170
  threat posed by, 172–175
Nemesis star, 23
Neptune, 12
Nero, 33
Neukum, Gerhard, 172
Newton, H. A., 68
Newton, Isaac:
  and Edmund Halley, 35–36
  Principia, 36

Olivier, Charles P., 67
Olmsted, Denison, 67
Oort, Ian, 13
Oort comet cloud, 13, 21, 22
Oppolzer, Theodor von, 72
Orionid meteor shower:
  comet Halley and, 75
  described, 69

PAHs. *See* Polycyclic aromatic hydrocarbons
Pallas (asteroid), 122, 125, 126
Pathfinder spacecraft, 110
Peary, Robert, 82
Perihelion (of comets), 47–48
Periodicity theory (comets), 22–23
Perseid meteor shower:
  comet Swift-Tuttle and, 50, 71, 72
  described, 69
  orbit of, 72
Peters, C. F. W., 72
Phobos, 134–135
Piazzi, Giuseppe, 119, 121
Pioneer spacecraft, 78
Pizarro, Francisco, 34
Planetesimals:
  defined, 85
  evolution of, 85–86, 121–122

Pluto:
nature of, 20–22
orbit of, 20
Poag, Wylie, 164
Pogson, Norman, 75
Polycyclic aromatic hydrocarbons, 112
Pons (comet). *See* Comet Pons
Pons, Jean-Louis, 42, 72
Popigai impact crater, 164–165
Poynting-Robertson effect, 131
Predictable comets:
comet Encke, 37–38
comet Halley, 35–37
Prograde rotation (meteoroids), 131

Quadrantid meteor shower:
comet 1491 I and, 75
described, 69

Raup, David, 23
Regolith, 86
Resonances (Jupiter and asteroids), 124
Retrograde rotation (meteoroids), 131
Rodari (asteroid), 59
Rosetta space probe:
comet Wirtanen and, 60–61
described, 60–61

Saturn, 12
Schiaparelli, Giovanni:
comet-meteor connection and, 70–72
Schiaparelli, Giovanni (*Cont.*):
Martian canals and, 70–71
Mercury and, 70
Notes and Reflections on the Astronomical Theory of Falling Stars and, 71
Schwassmann-Wachmann (comet). *See* Comet Schwassmann-Wachmann
Scotti, Jim, 169, 170
Secchi, Angelo, 70
Seki, Tsutomu, 48
Shakespeare, William, 35
Shergotty meteorite, 99
Shoemaker, Carolyn, 51
Shoemaker, Gene, 51, 56, 172
Shoemaker-Levy 9 (comet). *See* Comet Shoemaker-Levy 9
Short-period comets, 20
Sikhote-Alin meteorite fall, 92
Skjellerup-Maristany (comet). *See* Comet Skjellerup-Maristany
Smithsonian Astrophysical Observatory, 42, 43
SNC meteorites:
evidence pointing to Mars as source, 101–103
See also Martian meteorites
SOHO-6 (comet). *See* Comet SOHO-6
Solar and Heliospheric Observatory (SOHO) spacecraft:
comet Hyakutake and, 31
comet SOHO-6 and, 48

Solar nebula, 83, 85
Solar system:
birth of, 11–13, 82–85, 121–122
scale of, 123
Solar wind, 19
Southern Comet, 27
Spaceguard Survey, 172
Spahr, Tim, 171
Spectra (comets), 17
Stardust, 83–84
Stardust space probe mission:
comet Wild 2 and, 58–59, 79
described, 42, 58–59
Star of Bethlehem, 32
Staton, Buford, 91
Steel, Duncan, 41
S-type asteroids, 127
Gaspra, 135–138
Ida, 138–142
Sumners, Carolyn, 31
Sun:
birth of, 11
effect on comets, 47–49
Sungrazing comets, 48
Swift-Tuttle (comet). *See* Comet Swift-Tuttle

Tallakoteah, 82
Taurid meteor shower, 50
Tears of Saint Lawrence. *See* Perseid meteor shower
Terrestrial impact craters. *See* Impact craters
Thatcher (comet). *See* Comet Thatcher 1861 I
Themis asteroid family, 129
Titius, Johann, 120
Titius and Bode rule (astronomical spacing), 120–121
Tombaugh, Clyde, 20–21
Toscanelli, Paolo, 34
Trojan asteroids, 124
Tsunamis, 173
Tunguska impact event, 6, 155–157, 172
Tuttle (comet). *See* Comet Tuttle
Twining, A.C., 67

U-2 spy plane, 77–78
Uranus, 12, 120
Ursid meteor shower:
comet Tuttle and, 75
described, 69

Velikovsky, Immanuel, 41
Venus:
impact craters on, 151
meteorites from, 114
Vesta (asteroid), 125
Viking spacecraft, 110, 111, 134
Voyager spacecraft, 78

War of the Worlds (Wells), 99
Water:
comets and, 13, 14
Mars and, 110–111
Moon and, 13
Weiss, Edmund, 74
Welles, Orson, 99
Wells, H. G., 99
Whipple, Fred, 15

Whipple shield, 58
Whiston, William, 40–41
White-Ortiz-Bolelli (comet). *See* Comet White-Ortiz-Bolelli
Wild (comet). *See* Comet Wild 2
Wilson-Hubbard (comet). *See* Comet Wilson-Hubbard
Wirtanen (comet). *See* Comet Wirtanen
Wu, King, 33

Yarkovsky effect, 131

Zagami meteorite, 100
Zodiacal light, 14, 65–66

ABOUT THE AUTHORS

Carolyn Sumners, Ph.D., is Director of Astronomy at the Houston Museum of Natural Science, one of the country's leading space museums. A frequent lecturer on cosmic collisions, Dr. Sumners is also a consultant to the Central Bureau for Astronomical Telegrams, the agency that tallies the orbits of asteroids and comets throughout the solar system.

Carlton Allen, Ph.D., is a planetary geologist with Lockheed Martin at the Johnson Space Center. His research includes the Moon's resources, the soil of Mars, and the search for life in extreme environments.